Neural Network Models
Theory and Projects

Springer

London
Berlin
Heidelberg
New York
Barcelona
Budapest
Hong Kong
Milan
Paris
Santa Clara
Singapore
Tokyo

Philippe De Wilde

Neural Network Models

Theory and Projects

Second Edition

With 84 Figures

 Springer

Philippe De Wilde
Department of Electrical and Electronic Engineering
Imperial College of Science, Technology and Medicine
Exhibition Road, London SW7 2BT, UK

ISBN 3-540-76129-2 2nd edition Springer-Verlag Berlin Heidelberg New York

ISBN 3-540-19995-0 1st edition Springer-Verlag Berlin Heidelberg New York

British Library Cataloguing in Publication Data
De Wilde, Philippe
  Neural network models : theory and projects. - 2nd ed.
  1.Neural networks (Computer science)
  I.Title
  006.3'2
ISBN 3540761292

Library of Congress Cataloging-in-Publication Data
A catalog record for this book is available from the Library of Congress

First published 1996
Second Edition 1997

Typesetting: Camera ready by author
Printed and bound at the Athenæum Press Ltd., Gateshead, Tyne and Wear
69/3830-543210 Printed on acid-free paper

Contents

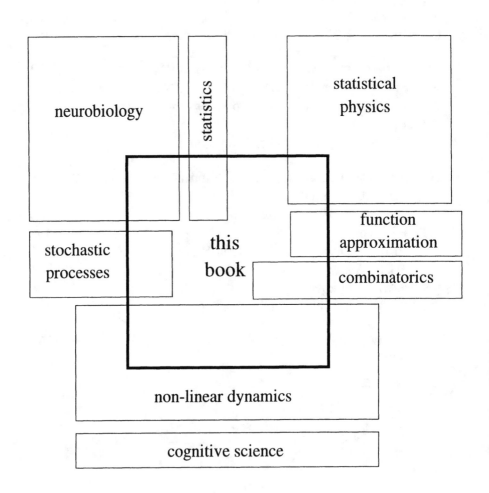

Preface

After a period of rapid growth, a successful enterprise enters a stage of consolidation. This is happening to the field of artificial neural networks. During the growth stage, there was a widespread cross fertilization between the different disciplines that contributed to neural networks. Nowadays, the field of neural networks consists of sub-fields, each related to a major branch of science. In the last nine years, I have done research into neural networks, while working and teaching in an engineering environment. I have learned that the knowledge needed to develop successful engineering applications of neural networks, is based on dynamical systems, part of information theory, and stochastic processes. This book uses these three paradigms to explain neural networks. The novelty of this approach is that dynamical systems and some information theory concepts are used for the first time in an expository text about neural networks.

This book explains and proves the main results in the dynamics of neural networks. It derives the non-linear dynamics from neurobiological principles. It investigates the stability and the convergence behaviour. Another important part is the capacity of discrete neural networks. Several capacity results are derived, using signal to noise ratios, and properties of threshold functions. I also introduce stochastic networks and simulated annealing. These are studied using concepts from Markov processes, rather than statistical physics.

Although I concentrate mainly on networks with feedback and their dynamical behaviour, one chapter is devoted to backpropagation, a popular technique in feedforward networks.

An important part of this book are the projects at the end of each chapter. Theory cannot be understood without practice, and practice for today's designing engineer or scientist means computer simulations. I have added eight challenging projects, one to each chapter. They are not straightforward applications of the material in the other sections, but they help the student develop an *integrated* knowledge of the theory, placing it within an application domain. All projects have ample introductions, and the reader's hand is held during the first part. Towards the end there are questions which could be developed into research. The projects require about two week's work in their basic form. This book can be used for a one-semester course.

This book is not about applications, but about techniques that can be used in applications. Giving examples of applications, as many books do, does not mean that one will be able to generalize the solution to a similar application. Teachers should give the student the necessary confidence to be able to apply neural networks techniques alongside others in any situation where pattern recognition is useful. How large the possible application range is, is shown at the end of chapter 2.

This is not a wordy book. It concentrates on essentials, but proves virtually every claim that is made. As any good design, it works through the usefulness of what it offers, not by its ornaments. It does not cover the feedforward neural networks that have developed into a branch of statistical classification. Books in this field are mentioned in the guide to the literature in chapter 1. Networks using radial basis functions need to be considered in a function approximation framework. Statistical mechanics and its glassy dynamics are not covered either, because this book is not a physics textbook. Noise is treated in the context of Markov processes. Synaptic dynamics are omitted, as are cognitive models.

I have taught a neural networks course at Imperial College of Science, Technology and Medicine, London, for several years now. My audience has included final year undergraduates, Master's degree students, and PhD students from Electrical Engineering, Computing, and Biomedical Systems. This book has been used in various forms as lecture notes and coursework text. It treats topics that are not readily available in other textbooks. It is also suitable for students and interested readers with a background in Applied Mathematics, Physics, and General Engineering.

Chapter 1 gives a bird's eye view of neural networks. It introduces all concepts that will be studied in depth in further chapters. The most popular neural networks are layered in structure, and they are studied in chapter 2. Neural networks are derived from models of the brain. This is the subject of Chapter 3. In this book, brain models are used as inspiration to build engineering systems, as the wings of a bird were the inspiration for the wings of an airplane. In chapter 4, we will see that neural networks are essentially dynamical systems, and we derive results about the stability of their equilibria. Chapter 5 makes the link between continuous and discrete neurons, and is also about oscillations in neural networks. The number of equilibria is calculated in chapters 6 and 7. Finally, chapter 8 introduces statistical models of neural networks.

It is assumed that the reader knows what a differential equation is. This is used in chapters 3, 4, and 5. For chapter 6, the reader needs to know what a binomial coefficient is, and what the equation of a line is. In chapter 7, the expression for the normal distribution is used. Chapter 8 uses Markov chains. *All chapters can be read independently of each other. The reader who does not like a chapter can just skip it and go on with the next chapter. Somebody in a great hurry can just read the first and the last chapter!*

I can be reached by electronic mail to p.dewilde@ic.ac.uk, and will be happy to answer any technical questions you may have about subjects treated in this book.

Many thanks to I. Aleksander, E. C. Van Der Meulen, colleagues at the Department of Electrical and Electronic Engineering of Imperial College, and at the Intelligent Systems Unit of BT Labs, Martlesham Heath. They have encouraged me an taught me practical things long after I graduated. In addition, the following people have inspired me via lectures or talks: D. Rumelhart, H. Haken, V. I. Kryukov, A. N. Chetaev, E. Caianiello, J. Eccles, E. Gelenbe, S. I. Amari, and J. Hopfield.

Thanks to all the staff at Springer-Verlag London Ltd. involved in the production of this book, especially Nicholas Pinfield and Michael Jones, for their encouragement and efficiency.

I dedicate this book to my wife Janet.

Philippe De Wilde

London
March 1997

Chapter 1

Key concepts in neural networks

This chapter gives a quick overview of neural networks. The key concepts are introduced, and the basic operation of the network is explained. Everything introduced in this chapter is taken up again later in the book.

There is also a guide to the literature, referring to some of the most inspiring books on neural networks. But this book stands on its own. In fact, it introduces and derives many results that you will not find in any of the existing books about neural networks.

After going through this chapter, you will be able to impress any audience at a cocktail party with your knowledge about neural networks!

1.1 Keywords and a Key Concept

Neural networks are about *associative memory* or *content-addressable* memory. You give a content to the network, and you get an address or an identification back. You could store images of people in the network. When you show an image of a person to the network, it will return the name of the person, or some other identification, e.g. the social security number.

Neural networks are also about *parallel processing* . They consist of a network of processors that operate in parallel. This means they

1

will operate very fast. To date, the most complex neural networks that operate in parallel consist of a few hundred neurons. But the technology is evolving fast. To recognize images, one needs about one processor per pixel. The processors, also called neurons, are very simple, so they can be kept small.

Neural networks are also *fail-soft devices*. If some of the processors fail, the network will degrade slowly. This is highly desirable from the consumer's point of view, and also for the design engineer, who has to cram many processors on one chip, and who can now use some less reliable, but smaller components.

Neural networks are often used to *classify* or *categorize*. If the images of a group of persons were stored in the network, it could infer who looks sad, and who looks happy. Of course, the only information the network has is the image, and someone could look sad, but still be very happy!

Neural networks are *distributed* systems. The knowledge the system has is distributed over the individual nodes, and their links. There is no central node that is controlling the other nodes. Every node takes individual decisions based on information it gets from the other nodes. This is a very natural and economical way of operating for large systems, obviating the need for expensive and difficult to maintain central databases. Some telecommunication networks, and the World Wide Web operate in this way.

The major key concept in neural networks is *the interaction between microscopic and macroscopic phenomena*. A macroscopic phenomenon of the image of a face of a person can be a frown, or a smile. A microscopic phenomenon is the correlation between the pixels in the image. Macroscopic phenomena are related to *form*. The spiral in figure 1.1 for example is generated by varying the curvature of a curve. When doing this, one only considers a few successive points on the curve. There is no global knowledge necessary to draw the spiral.

Growth of a crystal is another example of how microscopic or molecular forces have a macroscopic effect. Clouds are yet another example. The cloud is a *large scale* structure. The water molecules in it are on a *small scale*.

Much of the research in neural networks is about explaining *global properties* from *interaction between particles*. A good model of this is the derivation of the gas laws from the movement of molecules. In the

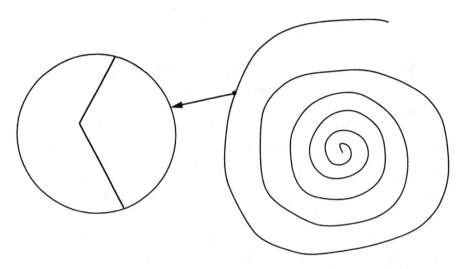

Figure 1.1: A spiral is a large scale form (right), generated by small scale interactions (magnified in the circle on the left) between the cells when the spiral is grown.

eighteenth century, Robert Boyle observed that for a given mass of a gas held at a constant temperature, the pressure is inversely proportional to the volume. This is a macroscopic property of a gas. At the end of the nineteenth century, Ludwig Boltzmann derived this relation using the speed and mass of the molecules in the gas. The molecules are the microscopic constituents of the gas.

For the network that recognizes faces of people, the macroscopic properties that we will study include the number of faces that the network can recognize, the speed at which it does this, the number of classes into which it can split up the faces (smiling, frowning, ...), etc. The microscopic interactions will be the calculations that the neurons, the little processors attached to each pixel, perform. We will have to specify these calculations so that the network can recognize a particular set of faces. Most applications of neural networks can be reduced to pattern recognition, as explained in section 2.6. So our example of recognizing a face turns out to be a very general one.

1.2   A Guide to the Literature

This little guide is deliberately selective, and presents the author's view on the most important neural network books and review papers.

This book is closest in spirit to J. Hertz et al., *Introduction to the Theory of Neural Computation* [56], and S. Haykin, *Neural Networks* [53]. They are the two main reference works for the theory of neural networks. They give a broad overview, and Hertz et al. also establish the roots of neural networks in statistical mechanics. Another excellent work, more encyclopedic in nature, is [12]. It is especially suited for the beginning researcher.

Several books present neural networks in a similar way, but are more specialized. Y. Kamp and M. Hasler, *Recursive Neural Networks for Associative Memory* [63] is good for Hopfield networks. E. Goles and S. Martinez, *Neural and Automata Networks* [45] stresses the link with cellular automata. S. Grossberg, *The adaptive brain* [50] studies biologically plausible neural networks as dynamical systems. The works [67, 74] study unsupervised classification. General iterative systems are studied in F. Robert, *Discrete Iterations* [93]. R. Hecht-Nielsen, *Neurocomputing* [55] is good for backpropagation, if one is tolerant of some peculiar notation. Some neural networks are statistical classifiers. They are reviewed in [17]. This work is useful to anyone using backpropagation.

Many theoretical results about neural networks stem from the statistical physics of spin glasses. A standard work is M. Mézard et al., *Spin Glass Theory and Beyond* [82]. The neural network side is presented in D. Amit, *Modeling Brain Function* [10].

A different approach, based on the Fokker-Planck equation, is taken by H. Haken in *Synergetics* [51]. This pioneering book predates neural networks themselves, and is a jewel to read. Other stochastic models of neural networks are studied in two largely unknown Russian books: A. Chetaev, *Neural Nets and Markov Chains* [21], and V. Kryukov et al., *The Metastable and Unstable States in the Brain* [71]. Some of Chetaev's work has been published in English as [20].

Two seminal papers by J. Hopfield are still good reading, and they have the advantage of brevity [60, 61].

Neural networks is an interdisciplinary field. Apart from statistical physics, it is also deeply embedded in cognitive science. When the link

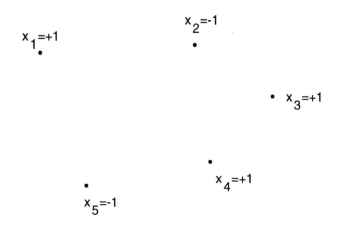

Figure 1.2: A network of five neurons, with their states indicated.

with cognitive science is stressed, neural networks are often referred to as *parallel distributed processing* or *connectionism*. Fundamental work in this field was done by a group of people around D. Rumelhart and J. McClelland. It is written up in their books about parallel distributed processing [94, 78, 77]. Another original approach, combining Boolean networks and cognitive science, is taken by I. Aleksander and H. Morton, in [9, 8]. All these books are easy to read, even without any knowledge about psychology or artificial intelligence. A link between neural networks and quantum mechanics is described in [37, 38, 31, 32].

On the Internet, `comp.ai.neural-nets` is an active newsgroup that has a regular posting answering frequently asked questions about neural nets. It is worthwhile reading this. It also contains information about simulation software.

1.3 The Operation of a Basic Network

1.3.1 Static Description of the Network

In figure 1.2, the neurons are simply represented as dots. To every neuron is attached a number that can be +1 or -1. This is called the *state* of the neuron. For example, the state of neuron 4 is denoted by x_4, and $x_4 = +1$. We will encounter neurons with more general states in section 1.4.4.

Figure 1.3: The network at time t=0.

The states of all neurons can be grouped in a vector, called the *state vector*. For the network in figure 1.2, the state vector is

$$\mathbf{x} = (+1, -1, +1, +1, -1)^T.$$

We will denote vectors by bold characters. The $\cdot^T$ indicates the transpose. This means that $\mathbf{x}$ is actually a column vector with 5 rows and 1 column. In general, the state vector has as many components, or rows, as there are neurons.

If the neurons were used to store pixels in the image of a face, a state +1 could indicate a black pixel, a state -1 could indicate a white one. Why +1 and -1 are used instead of 1 and 0 is a deep problem that will be discussed in section 7.2. For the moment, it is sufficient to remark that the numbers +1 and -1 can be transformed to arbitrary numbers a and b via the transformation

$$y = \frac{a(1 - x) + b(1 + x)}{2}.$$

1.3.2   Evolution in Time

Neural networks are dynamical. This means that their state changes in time. In figure 1.3, the network is drawn at time t=0. The state vector at this time is denoted by $\mathbf{x}(0)$.

In figure 1.4, the network is drawn at time t=1. Remark that neuron 1 has not changed state, but neuron 2 has changed state from -1 to +1. The state vector at time t=1 is denoted by $\mathbf{x}(1)$. Do not

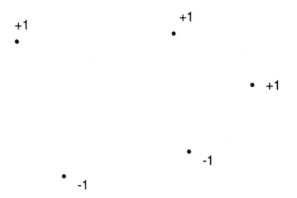

Figure 1.4: The network at time t=1.

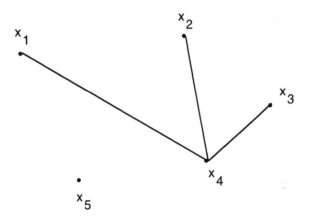

Figure 1.5: How neurons 1, 2, and 3 influence neuron 4.

confuse this with x_1, the state of neuron 1. As time evolves further, $t = 0, 1, 2, \ldots$, the state vector evolves too, $\mathbf{x}(0), \mathbf{x}(1), \mathbf{x}(2), \ldots$.

1.3.3 Construction of the Update Rule

The rule according to which the neurons change in time is called the *update rule*. Let's concentrate on neuron 4. In figure 1.5, you can see that there are links between neurons 1, 2, 3, and 4. This means that neuron 4 will be influenced by neurons 1, 2, and 3. This happens according to the following rule

$$x_4(t + 1) = \text{sgn}(T_{41}x_1(t) + T_{42}x_2(t) + T_{43}x_3(t)).$$

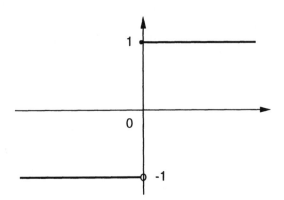

Figure 1.6: The function sgn. The value at 0 equals 1.

This formula shows how the state of neuron 4 at time t+1 is derived from the states of neurons 1, 2, and 3 at time t. The T's can be any number, and the function sgn is drawn in figure 1.6.

If we substitute some actual values in the update rule,

$$x_4(1) = \mathrm{sgn}[2(+1) + 2(-1) - 3(+1)] = \mathrm{sgn}(-3) = -1,$$

indicating that the state of neuron 4 at time t=1 is -1.

To understand the meaning of the numbers T, assume that neuron 4 is only connected to neuron 1. The update rule is then

$$x_4(t + 1) = \mathrm{sgn}(T_{41}x_1(t)).$$

If T_{41} is positive, this further simplifies to $x_4(t+1) = x_1(t)$. This means that neuron 4 will switch to the same state as neuron 1. In this case, the connection between neurons 4 and 1 is called *excitatory*. If T_{41} is negative, the update rule becomes $x_4(t+1) = -x_1(t)$, and neuron 4 will be in the state with opposite sign from the state of neuron 1. In this case, the connection is said to be *inhibitory*. The biological motivation for the terms excitatory and inhibitory will be explained in chapter 3. The elements of the matrix T are called *weights* or *synapses*.

In figure 1.7, all connections for a particular five neuron network are drawn. The update rule for a general neuron i in this network can be written

$$x_i(t + 1) = \mathrm{sgn}\left(\sum_{j=1}^{5} T_{ij}x_j(t)\right).$$

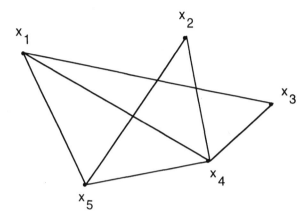

Figure 1.7: Some connections that can occur in a network of 5 neurons.

This formula can be repeated for $i = 1, 2, 3, 4, 5$.

The matrix T is called the interconnection matrix, or *synaptic matrix*. Synaptic is another term from biology that will be explained in chapter 3. When there is no connection, the corresponding element in the matrix T is zero. For the network in figure 1.7,

$$T = \begin{pmatrix} 0 & 0 & \cdot & \cdot & \cdot \\ 0 & 0 & 0 & \cdot & \cdot \\ \cdot & 0 & 0 & \cdot & 0 \\ \cdot & \cdot & \cdot & 0 & \cdot \\ \cdot & \cdot & 0 & \cdot & 0 \end{pmatrix}.$$

Remark that T is symmetric, and has zero diagonal. This means that neuron i influences neuron j in the same way that neuron j influences neuron i (T symmetric). Also, no neuron influences itself (zero diagonal).

For a network with n neurons, the update rule can be written

$$x_i(t+1) = \text{sgn}\left(\sum_{j=1}^{n} T_{ij} x_j(t) \right), \ i = 1, 2, \ldots, n. \tag{1.1}$$

If every neuron calculates its new state at the same time, the update is called *synchronous*. If, at every time step t= 1, 2, ..., only one neuron calculates its new state, the update is called *asynchronous*.

Figure 1.8: Convergence to a single state.

Problems of synchronous and asynchronous updating are discussed in chapter 5.

1.3.4   Trajectories in State Space

Following the evolution of a state vector through time, you can obtain something like (remember that $\cdot^T$ stand for the transpose of the vector, it makes a row vector into a column vector)

$$
\begin{aligned}
\mathbf{x}(0) &= (+1-1+1+1-1)^T, \\
\downarrow \qquad & \qquad\qquad \downarrow \\
\mathbf{x}(1) &= (+1+1+1+1-1)^T, \\
\downarrow \qquad & \qquad\qquad \downarrow \\
\mathbf{x}(2) &= (+1+1+1-1-1)^T, \\
\downarrow \qquad & \qquad\qquad \downarrow \\
\mathbf{x}(3) &= (+1+1+1-1-1)^T. \\
\downarrow \qquad & \qquad\qquad \downarrow
\end{aligned}
$$

After t=2, the state vector does not change any more. The network has *converged*. This is illustrated in figure 1.8.

For the network that stores faces, $\mathbf{x}(0)$ could be a digitized face from a security camera, with low resolution and noise. The network would then converge to one of the faces previously stored, for example from a database of criminals.

The network does not always converge, for example

$$
\begin{aligned}
x(0) &= (+1+1+1-1-1)^T, \\
\downarrow \qquad & \qquad\qquad \downarrow \\
x(1) &= (-1-1-1-1+1)^T, \\
\downarrow \qquad & \qquad\qquad \downarrow \\
x(2) &= (+1+1+1-1-1)^T, \\
\downarrow \qquad & \qquad\qquad \downarrow \\
x(3) &= (-1-1-1-1+1)^T, \\
\downarrow \qquad & \qquad\qquad \downarrow
\end{aligned}
$$

Figure 1.9: Oscillation between two states.

This network is *oscillating*, see figure 1.9. The network hesitates between two states, just as a human can hesitate between two persons, when recalling a face. Convergence and oscillation will be studied in chapter 4 and 5.

The sequence of states, as time evolves, is called a *trajectory*. The trajectory is a path in a higher dimensional space. Because the state vector has as many components as there are neurons, the higher dimensional space has as many dimensions as there are neurons. This space is called the *state space*. The components of the state vectors are either +1 or -1, and we will denote the state space by $\{-1,1\}^n$. This is for n neurons. The state space consists of the corners of an n-dimensional hypercube.

An example of a state space for a network with 5 neurons is sketched in figure 1.10. There are $2^5 = 32$ states. The end points of the trajectories are called *fundamental memories* or *attractors*. All states that are on trajectories going to the same fundamental memory are said to be in the same *attraction basin*.

The network can be set up to perform *error correction* in the following way: the fundamental memories are the pure images. The other points on the trajectories are the noisy or distorted images. As time evolves, the state of the network will converge to one of the pure images, or oscillate between a number of images. The starting state or initial state is sometimes called a *probe*.

1.3.5 Capacity

The *capacity* of a neural network is the number of fundamental memories that it can have. Suppose a user wants to store m fundamental memories

$$\mathbf{x}^1, \mathbf{x}^2, \dots, \mathbf{x}^m.$$

The i-th component of fundamental memory $\mathbf{x}^\alpha$ will be denoted by x_i^α. The following recipe for the synaptic matrix T was proposed by

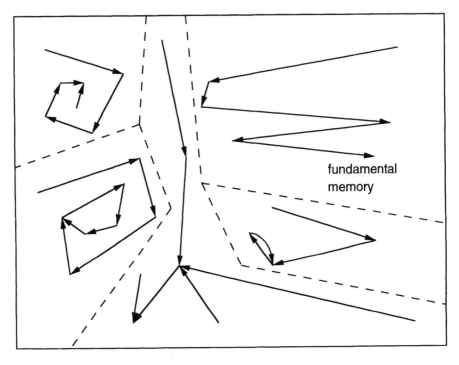

Figure 1.10: An example of the trajectories in the state space of a 5 neuron network.

J. Hopfield [60]

$$T_{ij} = \sum_{\alpha=1}^{m} (x_i^\alpha x_j^\alpha - \delta_{ij}), \quad i = 1, \ldots, n, \quad j = 1, \ldots, n. \qquad (1.2)$$

We will call a network with this choice of weights, a *Hopfield network*. We will prove in chapter 7 that, if you use this recipe, the capacity C_1 is of the order of

$$C_1 = \mathcal{O}\left(\frac{n}{4 \log n}\right),$$

where $\mathcal{O}$ means the following: $f(n) = \mathcal{O}(g(n))$ if there exists a number n_1 such that for all $n > n_1, f(n) \le g(n)$. For many practical applications, n_1 can be around 30.

But in chapter 6, we will prove that, if T is arbitrary, with zero diagonal, for another capacity C_2,

$$C_2 \le n.$$

So, capacity is a subtle concept, and we will have to define it carefully. For example for four neurons, $n = 4$, the synaptic matrix

$$T = \begin{pmatrix} 0 & -1 & -1 & -1 \\ -1 & 0 & -1 & -1 \\ -1 & -1 & 0 & -1 \\ -1 & -1 & -1 & 0 \end{pmatrix}$$

allows 6 fundamental memories, and not 4, as would be expected from the capacity result. Indeed, it is easy to verify that each of the vectors

$$(+1 + 1 - 1 - 1),$$

$$(-1 - 1 + 1 + 1),$$

$$(-1 + 1 - 1 + 1),$$

$$(+1 - 1 + 1 - 1),$$

$$(+1 - 1 - 1 + 1),$$

$$(-1 + 1 + 1 - 1),$$

is a solution of

$$\mathbf{x} = \text{sgn}(T\mathbf{x}).$$

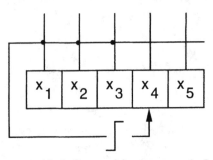

Figure 1.11: How neuron 4 is influenced by neurons 1, 2, and 3.

This is apparently in contradiction with the result $C_2 \leq n$, and we will solve this paradox in chapter 6.

The number of fundamental memories in a network is at most linear in the number of neurons. The fundamental memories are prototypes of classes. The other members in the class are slightly distorted versions of the prototype. During operation of the network, the trajectories lead from distorted inputs to prototypes. The same happens in a decoder for a communication channel. The prototypes are the codewords. During decoding a distorted codeword is associated with the nearest codeword.

Could a decoder replace a neural network? The number of codewords is usually exponential in the number of bits per word, and this is much better than a neural network, which has only a linear number of fundamental memories. The advantage is, however, that the neural network is completely parallel, and acting as an associative memory.

1.3.6   Auto-associativity

The mode of operation of the network, as we have studied it up to now, is sometimes called *auto-associative*. The network associates a starting state vector $\mathbf{x}(0)$ with a fundamental memory, after convergence. The fundamental memory is a vector with as many components as the starting state vector.

In figure 1.11, a schematic wiring diagram is shown for the connections between neurons 1, 2, 3, and 4. A dot on the crossing of two wires indicates a connection. Every square with x_i stands for a one-bit register that holds the state of neuron i. The update rule wired in the circuit of figure 1.11 is

$$x_4(t+1) = \text{sgn}(T_{41}x_1(t) + T_{42}x_2(t) + T_{43}x_3(t)).$$

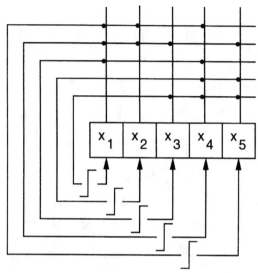

Figure 1.12: The wires in a network with 5 neurons. A dot on a crossing of two wires indicates that the weight is non-zero.

The general update rule

$$x_i(t + 1) = \text{sgn} \left(\sum_{j=1}^{5} T_{ij} x_j(t) \right)$$

can be wired as shown in figure 1.12.

1.4 Other Notions

1.4.1 Hetero-associativity (Associative Memory)

What is sometimes called *hetero-associativity* in a neural network context, is the real associative memory: associating an address with a content. In order to do this, nothing has to be changed to the auto-associative neural network, only the *interpretation* of the neurons changes. The network drawn in figure 1.13 uses neurons 1 and 2 for the address, and neurons 3, 4, and 5 for the content.

Because the address is unknown, the network is started with random values for the address, and a known content. The evolution in time of the state vector could be as follows:

$$(-1 - 1 + 1 - 1 + 1)^T,$$

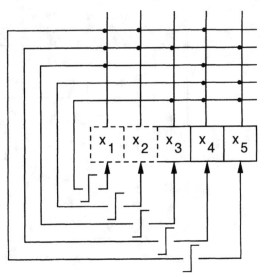

Figure 1.13: Neurons 1 and 2 represent the address. Neurons 3, 4, and 5 represent the contents.

$$(-1 + 1 + 1 - 1 + 1)^T,$$
$$(+1 - 1 + 1 - 1 - 1)^T,$$
$$(+1 - 1 + 1 - 1 + 1)^T.$$

The network retrieves the address (here +1 -1) corresponding to the contents (+1,-1,+1). Remark that you must allow the contents to change as time evolves! If the weights are well chosen, the network will not only retrieve the address, but even filter noise out of the data.

1.4.2 Layered Networks

Up to now, the input to the network was the state vector at time t=0, and the output was the state vector after convergence. There exist other ways to consider neural networks as black boxes that convert an input into an output. As with hetero-associative networks, nothing essential has to change, only the interpretation of the neurons. In the network in figure 1.14 for example, neurons 1 to 4 are the input, neurons 8 and 9 are the output. The other neurons are hidden in the black box, so we call them hidden neurons.

In many neural networks, the only connections allowed are from input neurons to hidden neurons, and from hidden neurons to output

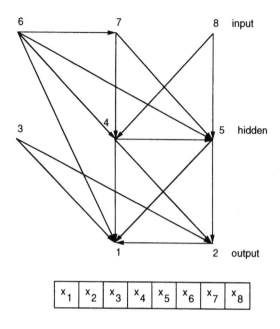

Figure 1.14: A very simple layered network.

neurons, as in figure 1.15. Such networks are called *multi-layer feedfor-ward networks*. The convergence in these networks is never a problem. First, the states of the hidden neurons are calculated, and then the states of the output neurons. Oscillations do not occur. We study the choice of the synapse matrix for these networks in chapter 2.

1.4.3 Learning in a Net with Input and Output

Multi-layer feedforward networks can *learn by example*. The examples are pairs of input and output, they are inputs for which the user knows the output. Learning consists in determining the weights, so that for every input, the network generates the corresponding output. This is also called *supervised learning*.

The weights are found by successively better approximations. First, a random guess is made. With these values of the weights, the output is calculated for a particular input. This output will be wrong, because the weights were guessed. Suppose that neurons 1 and 2 are the output neurons. They will have converged at a time $t = t_s$. Output neuron 1 will be in a state $x_1(t_s)$, whereas the correct state would be y_1.

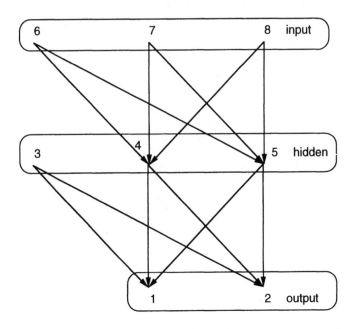

Figure 1.15: A layered network where the only connections are between successive layers.

Similarly for neuron 2. We can now define an error E as

$$E = [x_1(t_s) - y_1]^2 + [x_2(t_s) - y_2]^2.$$

This is a nonlinear function, dependent on the input to the network, or its starting state $\mathbf{x}(0)$, and the weight or synaptic matrix T:

$$E = E(\mathbf{x}(0), T).$$

When the network has to learn several examples, the error E can be summed over all examples, or pairs of input and output. This is further explained in chapter 2.

   The function E is nonlinear in T, because the neurons perform the nonlinear function sgn. Any algorithm for nonlinear function minimization can now be used to calculate the weights T. For this reason learning in a neural network is essentially minimization of a nonlinear function.

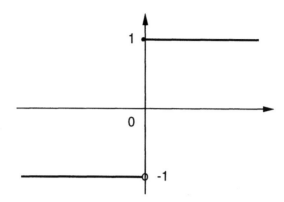

Figure 1.16: The function sgn. The value at 0 equals 1.

1.4.4 Other Update Rules

We have already mentioned that the error E for learning is nonlinear because of the use of the function sgn. In figure 1.16 we define this function, and also indicate what happens for zero input. Zero input to neurons is a tricky situation, and we discuss it in chapter 5.

If the sgn function is used in the neurons, the states of the network are on the corners of a hypercube,

$$\mathbf{x}(t) \in \{-1, +1\}^n.$$

Sometimes, a threshold θ_i for neuron i is included, and the update rule becomes

$$x_i(t+1) = \text{sgn}\left(\sum_{j=1}^n T_{ij}x_j(t) - \theta_i\right), \quad i = 1, 2, \ldots, n. \qquad (1.3)$$

More general nonlinear functions can be used. It is usually required that they saturate at -1 and +1. An example is the function tanh, drawn in figure 1.17. The states are now in the interior of the n-dimensional hypercube,

$$\mathbf{x}(t) \in \,]-1, +1[^n.$$

For a general function f, the update rule is

$$x_i(t+1) = f\left(\sum_j T_{ij}x_j(t)\right), \quad i = 1, \ldots, n.$$

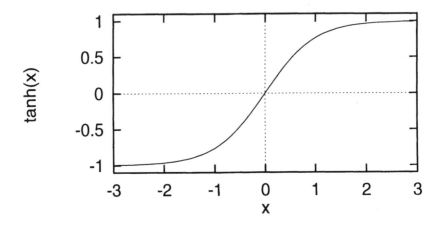

Figure 1.17: The function tanh.

Another possibility is to have the neurons operate probabilistically. Their state is again +1 or -1, but now only a probability is given for the state to be +1 or -1. This probability will be a function of the input to the neuron. For example,

$$\Pr(x_i(t+1) = 1) = g\left(\sum_j T_{ij}x_j(t)\right),$$

$$\Pr(x_i(t+1) = -1) = 1 - \Pr(x_i(t+1) = 1),$$

with g a function similar to tanh, but with values between 0 and +1 instead of -1 and +1. A possible choice is $(1 + \tanh)/2$.

The slope of the function g can be interpreted as a temperature in a physical system. This will be discussed in chapter 8. For a small slope, the neural network becomes chaotic. For a large slope, it becomes deterministic.

1.4.5 Energy Surface

It is possible to associate an energy function with the neural network. As time evolves, and the state of the network evolves, the energy will decrease. This is similar to a ball rolling down a slope: its potential energy decreases. A candidate for the energy function is

$$H(x) = k_1 \sum_{i,j=1}^{n} T_{ij}x_i x_j,$$

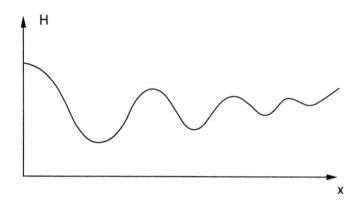

Figure 1.18: An energy function H with local minima.

with k_1 an arbitrary constant. This energy function is quadratic in the states of the neurons.

Often the energy function has local minima. Many of these correspond to unwanted fundamental memories, they are sometimes called *spurious* . An example is depicted in figure 1.18. We will study energy functions in chapters 4, 5, and 8.

1.4.6 Higher Order Networks

Up to now, the input to a neuron was linear in the states of the other neurons. There is no need for this limitation, and one can consider *second order networks*, where the input is quadratic in the states of the other neurons,

$$x_i = \text{sgn} \left(\sum_{j,k=1}^{n} T_{ijk} x_j x_k \right).$$

The weights T_{ijk} are now elements of a tensor, a higher dimensional matrix. There are no straight lines on the graph of the network anymore, as you can see in figure 1.19.

One could calculate the products of states of neurons in intermediate neurons. This is shown in figure 1.20. $T_{ijklm} x_j x_k x_l x_m$ has to be calculated. On the two branchpoints in the graph, intermediate neurons could calculate $x_j x_k$ and $x_l x_m$.

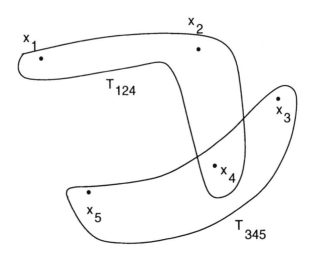

Figure 1.19: A network with second order interactions between neurons 1, 2, and 4, and between 3, 4, and 5.

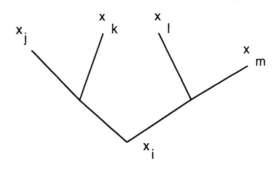

Figure 1.20: A network with fourth order interactions between neurons j, k, l, m, and i

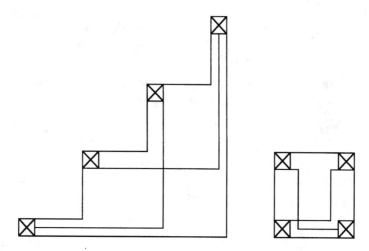

Figure 1.21: Two schematic VLSI layouts of four neurons. Every neuron is drawn as a square with a cross.

For a general higher order neural network, the update rule is

$$x_i(t+1) = sgn \left(\sum_{j_1,j_2,\ldots,j_d=1}^{n} T_{ij_1j_2\ldots j_d} x_{j_1}(t) x_{j_2}(t) \ldots x_{j_d}(t) - \theta_i \right),$$
$$i = 1,\ldots,n, \tag{1.4}$$

and the generalized sum-of-outerproducts rule for the weights is

$$T_{ij_1j_2\ldots j_d} = \sum_{\alpha=1}^{m} \left(x_i^\alpha x_{j_1}^\alpha x_{j_2}^\alpha \ldots x_{j_d}^\alpha - \delta_{ij_1j_2\ldots j_d} \right),$$

where $\delta_{ij_1j_2\ldots j_d}$ is zero if none of the subscripts is equal to another, and it is one otherwise.

1.4.7 Layout

As neurons are processors, the aim of the engineer will be to have as many as possible of them together on a chip. If one attempts this, it turns out that the connections between the neurons take up more chip area than the neurons themselves. The total length of the connections will depend on the placement of the neurons. In figure 1.21, two possibilities are shown. One has four neurons placed diagonally, the other has them placed as close together as possible. The latter placement will have a smaller total interconnection length, but there is a trade-off. If the neurons are placed on equally spaced grid points, one needs *channels* between the neurons to route the connections. As the figure shows, the closer the neurons, the denser the connections in the channels between the neurons. For n neurons, this trade-off can be seen in the following table.

total interconnection length	$\frac{(n+1)n(n-1)}{3}$	$\frac{\sqrt{n}n(n-1)}{3}$
sum of channel widths	$\frac{n(n-1)}{2}$	$\frac{\sqrt{n}n(n-1)}{6}$
maximal channel width	$n-1$	$\frac{n^2}{4}$

1.5   List of Symbols

Symbol	Meaning	Chap.
α, β	dummy index over patterns or states	all
γ_i	state of neuron, random variable, $+1$ or 0	7,8
i_i	state of neuron, between -1 and $+1$, layered net	2
$I(\mathbf{y})$	state of neuron at location $\mathbf{y}$	5
i, j, k	dummy index over neurons	all
ln	logarithm in base e	all
m	number of stored patterns	all
n	number of neurons	all
p_α	occupation probability	8
$Q_{\alpha\beta}$	transition probability	8
$\rho(\mathbf{y})$	density of neurons at $\mathbf{y}$	5
t	time	all
T_{ij}	weight in net with feedback	all
θ_i	threshold	all
u_i	neuron state, between -1 and $+1$, feedback net	3,4,5
v_i	state of neuron, $+1$ or 0, net with feedback	7
w_{ij}	weight in layered net	2
$w(\mathbf{y}, \mathbf{y}')$	weight from $\mathbf{y}'$ to $\mathbf{y}$	5
x_i	state of neuron, $+1$ or -1, net with feedback	5,6,7,8
ξ_i	state of neuron, random variable, $+1$ or -1	7,8

Sometimes we will use a different symbol for the same variable, to indicate that the network studied is fundamentally different. Within the same chapter, the symbols are always consistent, and in the whole book, no symbol is ever used for two different variables.

1.6 Problems

1. Is the state space for a neural network with input and output different from the state space for an auto-associative neural network?

2. Are there as many trajectories as fundamental memories in state space?

3. Compare the capacity of a neural net used as an associative memory with that of a normal memory consisting of several 32-bit words.

4. Does the state space of a dynamic, fully interconnected, neural net always consist of the corners of a hypercube?

5. Can a Hopfield net be considered a network with input and output?

6. Is it possible, with the current technology, to simulate a completely connected network of one million neurons?

1.7 Project: Semantic Networks

This chapter has shown that neural networks are essentially pattern recognizers. The input is a bitstring, the output is another bitstring that somehow classifies the input bitstring, either by showing a previously stored prototype, or by giving an address for the input.

In many applications, the state of a neuron represents the black or white value of a pixel in an image. This is straightforward pattern recognition, the neurons represent pixels. Sometimes the image is preprocessed, for example by taking a Fourier transform. In this case the neuron states do not correspond directly to pixel values.

Pattern recognition is however a concept much broader than mere classification of pixel arrays. This project aims at giving you some feeling of what a network of neurons can model.

At first, you may feel uncomfortable with the examples given. This is because they do not use a representation of data that you are used to. Most algorithms are developed for computers that operate sequentially. This implicitly causes a sequential organization of the data. We will use a network of processors operating in parallel. The representation of the data we will use is sometimes called connectionist, or distributed over the network.

1.7.1   Features Identifying a Room

Classifying a room from its features was an early success of neural networks [94], and is still a topic of current research [69].

Look at the room around you, and observe its features. Does it have windows? Is the ceiling high or low? How many tables are there? Is it bright? Is there a bed? Is there a bath? Are there other people in it? Identify twelve features that can occur in any sort of room you can think of.

Now look again at the room you are sitting in, and give each of the twelve features a score out of ten. A ceiling 7 meters high may give a score 10/10 for the height of ceiling feature, 2.5 meters high may give 5/10, 1 meter high may give 1/10. Remark that the scores are subjective and nonlinear. That's life. Similarly, assign scores for the other features. The scores for the room I am sitting in look like

+	+	+	+	-	-	-	-	-	-	height of ceiling
+	-	-	-	-	-	-	-	-	-	doors
+	+	+	-	-	-	-	-	-	-	windows
+	-	-	-	-	-	-	-	-	-	tables
-	-	-	-	-	-	-	-	-	-	bath
-	-	-	-	-	-	-	-	-	-	beds
+	+	+	+	+	+	-	-	-	-	brightness
+	+	-	-	-	-	-	-	-	-	people
+	-	-	-	-	-	-	-	-	-	animals
+	-	-	-	-	-	-	-	-	-	food
-	-	-	-	-	-	-	-	-	-	cars
-	-	-	-	-	-	-	-	-	-	rats

Remark that the scores are not numbers, but sequences of the sign +. Make similar profiles for the rooms that you frequent regularly.

We will now transfer the knowledge about the rooms into a neural network. Choose a network with 120 neurons. Every room profile will be one fundamental memory, + corresponding with a state +1 for the neuron, − with a state −1. The first ten neurons can represent the first feature, and so on. You have now as many fundamental memories as there are room prototypes.

The weights can be calculated with formula (1.2), where x_i^α is the state of neuron i in prototype α. Although you can write the prototypes down by hand, you will need a computer to calculate the weights.

Program the formula (1.2) yourself, it is the only way to get insight in the calculations. There exist many commercial neural network software packages. They are useful once you understand neural networks, because they simplify calculations, and you can use your insight to verify the results. At this stage you don't have this insight, and a commercial package could give you wrong results when you don't realize it. One of the most popular packages, for example, implements the rule (1.2) in such a way that the network has very few fundamental memories. Only after the study of chapter 4 would you be able to understand why.

Once you have calculated the 7140 weights, you can now use the network. Ask a friend to imagine a room without telling you, but to give you the profile of the room, consisting of the scores of the twelve features. Feed these scores into the computer as a vector with 120 components. This will be the state of the network at $t = 0$. Now apply the update rule (1.1). After some time, the network will be in a state corresponding to one of the prototype rooms that you programmed in it via the weights. This way, the network recognizes a room from its features.

Occasionally, the network may not converge to a known prototype. It may have converged to a state which is equal to a prototype, but with all +1 and −1 interchanged. You should have no problem to recognize the prototype from its inverted version. Another possibility is that the network converges to a mixture of two prototypes. This can often be avoided by increasing the number of neurons. Exactly how many new neurons you need will be clear after chapter 7.

Play around with your network for a while, adding some new prototypes, and presenting it with a whole range of initial states. After a while, you should be able to answer the following questions.

1. What is easiest to add, prototypes or features?

2. In what way can you reuse the old weights when you add a new prototype?

3. How long does the network on average take to converge? Does this depend on the number of prototypes, on the number of features?

4. Could you assign the twelve first bits of the features to the first twelve neurons, the twelve second bits to the next twelve neurons, and so on?

5. Try to replace the 10 neurons per feature by one neuron with 10 different states. By what function would you replace sgn in (1.1)? If you program this, do not try to store more than 2 prototypes.

6. Can you make the network identify the prototype?

1.7.2   Correlations and a Semantic Net

A dictionary explains words in terms of other words. You could write all words on a big page, and draw a line between two words if one of them was used in the explanation of the other. If you did this, you would see many clusters. These clusters of related words could correspond to sections in a thesaurus.

This principle has been used for some time in artificial intelligence. Several variants have been given different names, but we will refer to such a network of clusters as a semantic network [96]. Neural networks can implement semantic networks in a very simple way.

If there are people in a room, it is very likely that there are also chairs in this room. As far as rooms are concerned, there is a high correlation between people and chairs. There is a low correlation between people in a room and cars in a room. You are not likely to stay long in a room full of cars, i.e. a garage. There is a correlation between rats and food, and many more. You can attach the correlations to the links between the words. Such correlations are subjective. That's the way language works. In figure 1.22 is an example of a semantic network with correlations on the links.

Program a small semantic network of your own. The words could be features of a room, and the weights can be equal to the correlations

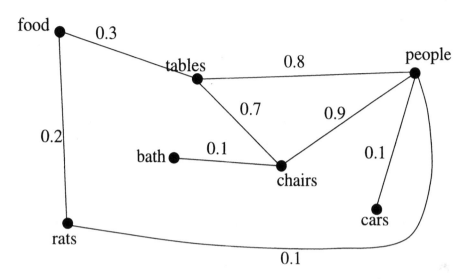

Figure 1.22: A small semantic network, indication correlations between features of a room.

between these features. Store the correlations in the weight matrix of a neural network.

What are the states of the neurons? They will be the degree to which you have identified a feature. You may not be quite certain that there are rats in a room. You may want to assign e.g. 0.6 as the state of the *rat* neuron. You can quantify your uncertainty as a number between 0 and 1. But the state of the neurons is between -1 and +1, so you need to multiply by 2 and subtract one.

Alternatively, you can take the scores of the features from the neural network in section 1.7.1, but rescale them from $[0, 10]$ to $[-1, +1]$. In this application, however, it is essential that you could not observe or measure all features. The input or the state of the network at time $t = 0$ will be an incomplete description of a room.

For the initial state of the neurons that correspond to unknown features, use random numbers between -1 and +1. Now that the initial state and the weights are known, we can apply the update rule (1.1), and let the network converge. The network will converge to a complete description of a room. The network converges again to a prototype, but unlike section 1.7.1, we did not store the prototypes, but we have stored the weights directly.

If you have programmed this, you can now answer the following questions.

1. Add nodes like *bathroom, garage, food locker* to the network. How would you calculate the weights of links that include these nodes?

2. If the network converges to a certain prototype, could you estimate how sure the network is of its decision? Is it possible to find how it reached the decision to identify a certain prototype?

3. Try a semantic network with weights in one direction only.

4. Can you express more complex relationships via weights on the links?

Semantic networks have a wide range of applications. Think about a few illnesses you have had, and their symptoms. Draw the graph, with the correlations between the symptoms. Other nodes will represent the diagnosis, the names of the illnesses. Fill in the correlations between the illnesses and their symptoms [42].

1.7.3   Rules and State Machines with Neural Networks.

You can easily implement Boolean functions , with the update rule (1.1) if you use e.g. +1 for TRUE, and -1 for FALSE. This is illustrated in figure 1.23 for NOT, AND, and OR. For AND and OR we have introduced neurons whose state is fixed in time, or clamped.

The networks for AND, OR, and NOT have two layers of neurons. Try to implement the exclusive-OR function in a similar way. You will need three layers of neurons.

A large part of human knowledge can be expressed as rules, using the basic Boolean functions AND, OR, and NOT. Try to make a neural network that implements the rule *D is true if and only if A and B are true or C is false.* If you use the basic building blocks from figure 1.23, you will find a network with three layers of neurons. Now, add some more rules to the same network, for example *E is true if and only if D and C are true*, and *A is false if and only if E or B are true.* Draw the resulting network.

While adding the last rule, you will have introduced feedback into the network, because you have defined some inputs into A, a neuron

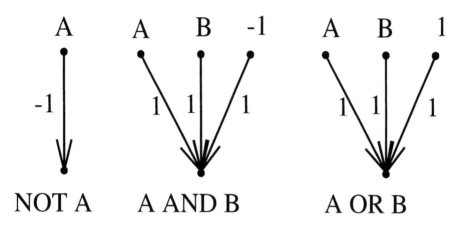

Figure 1.23: Neurons implementing the Boolean functions NOT, AND, and OR. The value TRUE is represented by +1, FALSE by -1. The weights are indicated, and also the state of clamped neurons.

whose state you have been using beforehand. The following has happened now. You have expressed part of your knowledge about the world in three rules. The truth values of B and C have to be specified. They are the external inputs into the system. The neural network will follow a trajectory in state space, depending on these inputs.

The state space has 32 points, representing all possible combinations of the variables A, B, C, D, and E being true or false. Draw the 32 points, and all possible trajectories.

You can simplify you task by choosing B and C first, and then drawing a state space with 8 points, corresponding to the combinations of A, D, and E. You have to do this four times, for all combinations of B and C.

What you have done is representing rule-based reasoning as trajectories in state space. This illustrates that it is possible to implement a rule-based expert system on a neural network.

In section 1.7.1 we have seen how a neural network recognizes patterns consisting of features. If you combine this pattern recognition with rule-based reasoning, you get a very powerful artificial intelligence system [42, 7, 70].

Chapter 2

Backpropagation

2.1 Introduction

This chapter describes how to calculate the weights or synapses of a multilayer feedforward network. The network will learn to associate a given output with a given input by adapting its weights. The weight adaptation algorithm we consider here is the steepest descent algorithm to minimize a nonlinear function. For neural networks, this is called *backpropagation* and was made popular in [78, 94].

In sections 2.1-2.4 we limit ourselves to fully interconnected neural networks with three layers. Some authors however, call those networks, networks with two layers, or networks with one layer. In section 2.5 we describe the general backpropagation network, for an arbitrary number of layers. Section 2.6 gives applications of backpropagation.

One of the advantages of the backpropagation algorithm, when implemented in parallel, is that it only uses the communication channels already used for the operation of the network itself. In this chapter, every effort has been made to keep the description of the algorithm independent (orthogonal) of the communication channels needed in a VLSI realization of it.

The algorithm presented incorporates a minimal amount of tuning parameters so that it can be used in most practical problems.

Backpropagation is really a simple algorithm, and it has been used in several forms well before it was "invented". As with most useful ideas, it is the result of the efforts of many scientists, some influencing each other, some working independently. It is an emergent property of

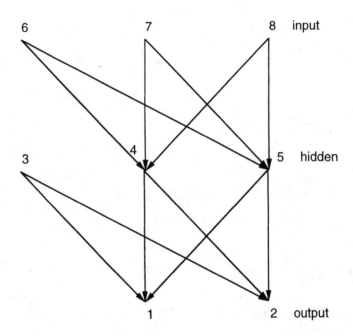

Figure 2.1: An example neural network with input neurons 7 and 8, hidden neurons 4 and 5 and output neurons 1 and 2. The state of neurons 6 and 3 is always 1.

the work of many, not the idea of a single individual.

2.2   How the Network Functions

Refer to figure 2.1 during the following description. The input is presented to neurons 7 and 8. The input to those neurons is either -1 or $+1$. The input can also be a continuous value in $[-1, +1]$, representing e.g. the grey value of a pixel or the uncertainty (multiplied by 2 and minus 1) of a statement. Each neuron will have a value associated with it. This value is called the *state* of the neuron. The state of neurons 7 and 8 is their input value. The state of neurons 6 and 3 will *always* be $+1$. We will say that these neurons are clamped to $+1$.

Each interconnection between two neurons has also a value associated with it. This value is called the *weight* of the interconnection. The weight of the interconnection between neurons 7 and 4 is denoted w_{47}. There is no weight w_{74}. The state i_4 of neuron 4 is now calculated as

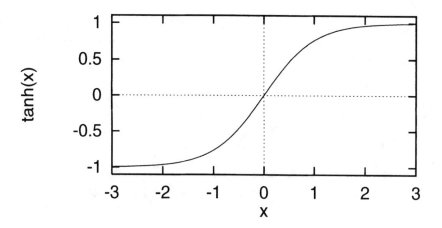

Figure 2.2: The function $\tanh x$.

follows.

$$i_4 = \tanh(w_{46}i_6 + w_{47}i_7 + w_{48}i_8).$$

Remember that $i_6 = +1$ in this expression. This is a way of providing *thresholds* w_{46} and w_{56} to neurons 4 and 5. The threshold shifts the input to the neurons. Thresholds also provide an increase in the dimensionality of the parameter space, and the more parameters a neural network has, the better it will be able to adapt.

We suppose that the weights are known. How to determine them is the subject of the next section.

The state of neuron 5 is calculated in a similar way as

$$i_5 = \tanh \sum_{j=6}^{8} w_{5j}i_j.$$

The function (see figure 2.2) $\tanh x$ is quite complicated to calculate, but the neural network will still work well with an approximation requiring only a few multiplications and additions.

Now the state of neurons 4, 5 and 3 is known (remember that $i_3 = +1$). This allows one to calculate the state of the output neurons 1 and 2:

$$i_1 = \tanh \sum_{j=3}^{5} w_{1j}i_j, \quad i_2 = \tanh \sum_{j=3}^{5} w_{2j}i_j. \tag{2.1}$$

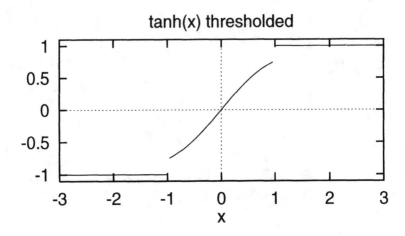

Figure 2.3: The thresholding procedure for the output neurons.

The states i_1 and i_2 will in general be numbers in $[-1, +1]$. But most of the time the neural network is used as a black box transforming an input string consisting of -1 and $+1$ into an output string consisting of -1 and $+1$. It is then necessary to interpret all numbers above e.g. 0.75 as $+1$, all numbers below -0.75 as -1, and all numbers in $[-0.75, +0.75]$ as *error* (see figure 2.3). The number 0.75 is the value of a parameter that we will call τ.

This ends the description of the operation of the neural network: in goes a string of -1 and $+1$ (one value per input neuron), out comes a string of -1, $+1$ and *error* signals (one value per output neuron). Up to now there were two inputs and two outputs, in section 2.4 a general three layer neural network will be described.

2.3 Calculation of Weights by Backpropagation

2.3.1 Calculation of the Error Signals of the Output Neurons

The user will want the neural network to mimic a particular input-output behaviour. For example, $i_7 = +1$ and $i_8 = -1$ should give $i_1 = -1$ and $i_2 = +1$. In order to obtain this, the weights will have to be adapted.

Start by making a random guess for the weights. Do not allow zero as a guess for a weight, it may stay stuck at zero during the backpropagation procedure. This is how a particular geometry can be imposed on the network. Offer $i_7 = +1$ and $i_8 = -1$ as an input to the neural network. Calculate the value for i_1, using the operation rules for the network described above, including the thresholding for the output neurons as described at the end of section 2.2. The value obtained for i_1 will in general be different from the expected or *target* value $i_1 = -1$. The difference between the target value and the value obtained will be the error signal e_1 that will be propagated back to the input neurons. In the same way the error signal e_2 for the output neuron 2 is calculated.

In short, the input-output pair is presented to the network and an error signal is accumulated in each of the output neurons.

2.3.2 Updates of Weights between Hidden and Output Neurons

The weights, initially chosen at random, will now be adapted in function of the error signals sent from the output neurons. The weight w_{14} will be changed by an amount

$$\Delta w_{14} = \eta e_1 (\mathrm{sech}^2 o_1) i_4.$$

In this equation o_1 is the state of neuron 1 *before* the application of the tanh function in formula (2.1):

$$o_1 = \sum_{j=3}^{5} w_{1j} i_j.$$

The function $\mathrm{sech}^2 x$ is the derivative of $\tanh x$. This derivative should not be too small. If it is nearly zero, one should consider adding a small positive number to it, see figure 2.4. An initial try might be to add 0.05.

The parameter η is called the learning rate, and has to be chosen by the user: a small η (typically 0.001) gives a slow changing of the weights, a large η (typically 10) might cause the weights to change by such large amounts that the network does not improve its performance in imitating the required input-output behaviour.

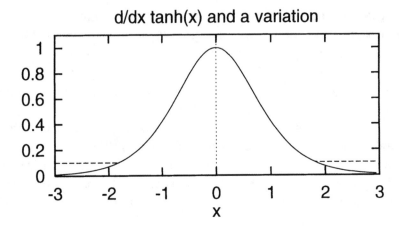

Figure 2.4: The altered derivative of $\tanh x$.

The following weight updates are obtained in a similar way:

$$\Delta w_{15} = \eta e_1 (\text{sech}^2 o_1) i_5,$$

$$\Delta w_{13} = \eta e_1 (\text{sech}^2 o_1) i_3,$$

$$\Delta w_{24} = \eta e_2 (\text{sech}^2 o_2) i_4,$$

$$\Delta w_{25} = \eta e_2 (\text{sech}^2 o_2) i_5,$$

$$\Delta w_{23} = \eta e_2 (\text{sech}^2 o_2) i_3.$$

2.3.3   Error Signals for Hidden Neurons

It is easy to assign an error signal to an output neuron, because there is an actual, calculated value for its state, and a target value. The error is then a measure of how far the actual behaviour is away from the target behaviour. This is not possible for the hidden neurons. Their error signal is derived from numerical analysis considerations. We just present the results here.

$$e_4 = \sum_{j=1}^{2} e_j (\text{sech}^2 o_j) w_{j4},$$

$$e_5 = \sum_{j=1}^{2} e_j (\text{sech}^2 o_j) w_{j5},$$

where o_j is defined in a similar way as o_1 in (2.3.2).

2.3.4 Updates of Weights between Input and Hidden Neurons

The updates are again determined from numerical analysis considerations. The results are

$$\Delta w_{47} = \eta e_4 (\text{sech}^2 o_4) i_7,$$

$$\Delta w_{48} = \eta e_4 (\text{sech}^2 o_4) i_8,$$

$$\Delta w_{46} = \eta e_4 (\text{sech}^2 o_4) i_6,$$

$$\Delta w_{57} = \eta e_5 (\text{sech}^2 o_5) i_7,$$

$$\Delta w_{58} = \eta e_5 (\text{sech}^2 o_5) i_8,$$

$$\Delta w_{56} = \eta e_5 (\text{sech}^2 o_5) i_6,$$

where i_7 and i_8 are the inputs to neurons 7 and 8. Remember that i_6, the input to neuron 6, is always $+1$.

2.3.5 Remarks on Epochs and Multiple Patterns

What has been described in the preceding subsections has the effect of changing all weights exactly once. This is called one *epoch*. It brings the network somewhat closer to imitating the required input-output behaviour.

In general several epochs will be needed before the network makes acceptable small errors. An error criterion can be the number of output neurons that gives a wrong output. One can also choose the sum of the error signals from the output neurons. The first error criterion is influenced by the choice of τ (see end of section 2.2).

In order to learn several patterns, show one pattern, adapt the weights for several epochs, then show another pattern, adapt again for several epochs, and so on until all patterns have been shown. Sometimes this whole process is called an epoch, rather than just one update of the weights. This method has a different error surface to be minimized for each pattern. It only works well if the minima are sufficiently high dimensional, i.e. the error function has a large plateau where it is minimal. This is one reason why there have to be a sufficient number of weights. The more weights, the larger the plateau will be, and the higher its dimension will be.

Another possibility is to present one pattern after another and ac-
cumulate the weight updates. The weight is updated after all patterns
have been shown to the network.

How many input-output pairs one can show to the network depends
on the number of neurons and the number of different weights, but the
dependence relation is not known and remains a major problem in
neural networks.

If the magnitude of the error keeps oscillating, it may be either
because an optimal set of weights has not been found, or because there
is no optimal set of weights. In the last case, the number of hidden
neurons may have to be increased.

Is it necessary to train noisy patterns? A related problem is whether
only to teach the network to recognize certain patterns, or also to reject
some. This is the subject of a whole branch of complexity theory. If
there are enough hidden neurons, it may not be necessary to teach the
network to reject certain patterns. If one has to, the patterns to be
rejected should not be too close to the patterns that are recognized.

2.4   A General Three Layer Network

The number of neurons in the input, hidden, and output layers is to
be chosen by the user. The number in input and output layers is often
fixed by the application. The determination of the minimum number of
hidden neurons needed is another unsolved problem in neural networks.
If the number of hidden neurons is very small, the network will classify
its inputs in a small number of classes. If it is large, the data will be
classified into many classes. How many classes are distinguishable in
the data is a problem from statistical data processing [35, 17].

In figure 2.5 is a sketch of a more general neural network than that
of figure 2.1. The equations for backpropagation easily generalize to
this case. It is difficult to write them down for an arbitrary multi-layer
network because the subscripts of the weights do not indicate between
which layers they are.

How many input-output pairs the network can learn is again an
unsolved problem. As a rule of thumb, do not try to learn more patterns
than 0.13 times the number of neurons.

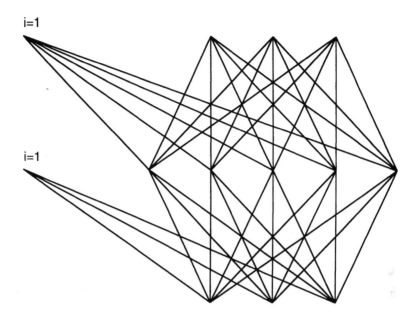

Figure 2.5: A more general neural network.

2.5 General Backpropagation Algorithm

Here we present the general backpropagation algorithm for a multi-layer network, but without any tuning parameters or neurons clamped at $+1$ or -1. We denote by L the number of *layers*. The different layers are denoted by $\mathcal{L}_1$ (the input layer), $\mathcal{L}_2, \ldots, \mathcal{L}_L$ (the output layer) (see figure 2.6).

We will assume that the output layer $\mathcal{L}_L$ contains M neurons. The neurons are distinguished by natural numbers. *The neurons in layer L are numbered from 1 to M.* When neurons n and j are connected, a *weight* w_{nj} is associated with the connection. In a *multi-layer feedforward* network, only neurons in subsequent layers can be connected:

$$w_{nj} \neq 0 \Rightarrow n \in \mathcal{L}_{l+1}, j \in \mathcal{L}_l, 1 \leq l \leq L - 1.$$

The total number of nonzero weights will be denoted by N.

What weights in the network are nonzero, i.e. the structure or geometry of the network, has been fixed in advance.

The *state* of the neuron j will be characterized by a real number i_j. The state of a layer, consisting of the states of all its neurons, will

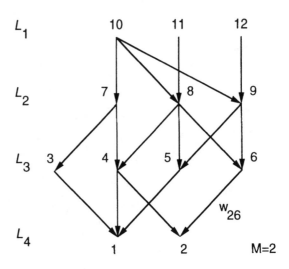

Figure 2.6: A multi-layer feedforward network

also be called a *pattern*. The network *operates* as follows. The state of the input layer is chosen, i.e. numbers i_j are assigned to the neurons $j \in \mathcal{L}_1$. Then the states of the neurons in layer $\mathcal{L}_2$ are calculated as follows:

$$i_n = f\left(\sum_{j \in \mathcal{L}_1} w_{nj} i_j\right), \quad n \in \mathcal{L}_2,$$

where f is a function that has a derivative for all possible values of its argument.

For a neuron n in an arbitrary layer l we will abbreviate by

$$o_n = \sum_{j \in \mathcal{L}_{l-1}} w_{nj} i_j, \quad 2 \le l \le L,$$

the sum of the signals it receives from the neurons it is connected to.

To determine the states of the third and subsequent layers, rule (2.5) will be applied sequentially: when the state of layer $l - 1$ is known,

$$i_n = f\left(\sum_{j \in \mathcal{L}_{l-1}} w_{nj} i_j\right), \quad n \in \mathcal{L}_l. \tag{2.2}$$

In order to *learn*, or to determine the weights, a number of pairs of states for the input and output layers are presented to the network.

We concentrate here on the choice of the weights when *one* pattern is presented to the network. The desired output pattern is described by the numbers t_m, $m \in \mathcal{L}_L$, $m = 1, \ldots, M$. Now suppose an initial guess of the weights w_{nj} has been made. Then the network can calculate the state of its output layer, i.e. the numbers i_m, $m \in \mathcal{L}_L$, $m = 1, \ldots, M$, according to rule(2.2). We now want to adapt our initial guess for the weights so that

$$E = \sum_{m \in \mathcal{L}_L} (t_m - f(o_m))^2 \tag{2.3}$$

is minimum.

If the function f in (2.2) is nonlinear, then the error E will be a nonlinear function of the weights w_{nj}. Any technique from nonlinear minimization can thus be applied to determine the weights. In neural networks the *gradient descent* (or steepest descent) method is commonly used. In this method the weights are updated proportional to the gradient of E with respect to the weights: the weight w_{nj} is changed by an amount

$$\Delta w_{nj} = -\eta \frac{\partial E}{\partial w_{nj}}, \quad n \in \mathcal{L}_l, j \in \mathcal{L}_{l-1}, \tag{2.4}$$

where the proportionality factor η is called the *learning rate*.

Let us calculate the updates for weights of connections between the last and next to last layer. Substituting (2.3) and (2.2) in (2.4), one obtains

$$
\begin{aligned}
\Delta w_{mj} &= -\eta \frac{\partial}{\partial w_{mj}} \sum_{p \in \mathcal{L}_L} \left[t_p - f\left(\sum_{q \in \mathcal{L}_{L-1}} w_{pq} i_q \right) \right]^2 \\
&= -\eta \, 2 \sum_{p \in \mathcal{L}_L} [t_p - f(o_p)](-f'(o_p)) \sum_{q \in \mathcal{L}_{L-1}} \delta_{pm} \delta_{qj} i_q \\
&= 2\eta [t_m - f(o_m)] f'(o_m) i_j, \quad m \in \mathcal{L}_L, j \in \mathcal{L}_{L-1}. \tag{2.5}
\end{aligned}
$$

The quantity $f(o_m)$ is present in the neuron m after the calculation of the output. We now assume that the neuron is not only able to calculate the function f, but also its derivative f'. Then the error signal $2\eta [t_m - f(o_m)] f'(o_m)$ can be sent back (or propagated back) to neuron j in layer $L - 1$. The value i_j is present in neuron j so that this neuron can now calculate Δw_{mj}. We will always suppose that the *weight w_{nj} is stored in neuron j.* In this way equation (2.5) describes the update of the weights (of connections) ending in layer $\mathcal{L}_L$.

In an analogous way it is possible to derive the updates of the weights ending in layer $\mathcal{L}_{L-1}$:

$$
\begin{aligned}
\Delta w_{nj} &= -\eta \frac{\partial}{\partial w_{nj}} \sum_{p \in \mathcal{L}_L} \{t_p - f[\sum_{q \in \mathcal{L}_{L-1}} w_{pq} f(\sum_{r \in \mathcal{L}_{L-2}} w_{qr} i_r)]\}^2 \\
&= -\eta\, 2 \sum_{p \in \mathcal{L}_L} [t_p - f(o_p)](-f'(o_p)) \sum_{q \in \mathcal{L}_{L-1}} w_{pq} f'(o_q) \times \\
&\qquad \sum_{r \in \mathcal{L}_{L-2}} \delta_{qn} \delta_{rj} i_r \\
&= 2\eta \sum_{p \in \mathcal{L}_L} [t_p - f(o_p)] f'(o_p) w_{pn} f'(o_n) i_j, \\
&\qquad n \in \mathcal{L}_{L-1}, j \in \mathcal{L}_{L-2}.
\end{aligned}
\tag{2.6}
$$

Now $2\eta \sum_{p \in \mathcal{L}_L}[t_p - f(o_p)] f'(o_p) w_{pn}$ is the weighted sum of the error signals sent from layer $\mathcal{L}_L$ to neuron n in layer $\mathcal{L}_{L-1}$. Neuron n can calculate this quantity because it has available the weights $w_{pn}, p \in \mathcal{L}_L$. It can then multiply this quantity by $f'(o_n)$ and send it to neuron j that can then calculate Δw_{nj} (and update w_{nj}).

This weight update can now be done for each layer in decreasing order of the layers, until $\Delta w_{nj}, n \in \mathcal{L}_2, j \in \mathcal{L}_1$ is calculated. This is the familiar backpropagation algorithm [56].

The backpropagation algorithm has all the convergence properties of the gradient descent algorithm. Although the advantages on the average seem to outweigh the disadvantages, one undesirable feature is the slow convergence when the gradient is small and the learning rate is fixed. More advanced algorithms can be used to replace the gradient descent. For an example, see [27, 28, 34].

2.6 Applications of Neural Networks

Here follow some applications of neural networks. Some require completely interconnected networks, not multi-layer ones, and so cannot be easily trained with backpropagation. But most will work well to some degree with multi-layer networks, trained with backpropagation. Some applications need hundreds of neurons. All the applications mentioned here have actually been demonstrated to work, the network being simulated in software. Between brackets is an estimate of the number of neurons one needs for a basic system. This estimate indicates an order of magnitude, e.g. 100 means: between 10 and 1000 neurons.

XOR (10) In the sixties, only networks without a hidden layer were considered. They could not implement the exclusive-OR function. It was therefore of major importance when it was realized that multilayer networks could implement this function. In fact, you can prove that, if there are enough hidden neurons, the network can implement *any* boolean function.

Encoder (10) If a parity bit is added in communicating a bitstring, an error during transmission can be detected. The input to the network would be the string without parity bit, the output would be the parity bit (one neuron). More general than a parity bit, an encoder performs a transformation of its input bitstring. Input will be the bitstring (as many neurons as there are bits), output the encoded string, usually longer than the input string.

Broomstick balancer (10) An important control problem. Balancing an inverted pendulum or broomstick is an essential task in the design of battlefield robots, for example. Input can be speed and acceleration of the pendulum, output the new position of the base of the inverted pendulum.

Speech synthesis (100) A major success of early backpropagation. A window of seven characters is slid over a text. The phonetic pronunciation of the middle character is produced. Seven input neurons, as many output neurons as there are phonetic characters. The hidden neurons take on a *meaning*: some will be active for vowels, others for consonants.

Human face recognition (100) One input neuron per pixel. Output neurons can identify the person, or grant or deny access to a room, etc. Similar techniques are used to identify somebody from an image of the retina of the eye, or an image of the hand.

Hyphenation algorithms (100) With 5 neurons one can encode $2^5 = 32$ characters. Several groups of 5 neurons can encode a word. The output can be the hyphenated word.

Backgammon (100) The ability to play games is important for management information systems. Input will be a board position. Number of input neurons is $\log_2$ of the number of different positions to be considered. Output is a move, suitably encoded.

Signal prediction and forecasting (100) This is part of many applications, from control of power stations to predicting exchange rates. The input neurons will represent previous values of the signal. The output will be a prediction of the future value. This is called a *tapped delay line*. The network is trained with inputs from the past for which a future value is known.

Recognizing handwritten ZIP codes (100) The input is a digitized image of a handwritten digit, one pixel per neuron. There can be ten output neurons, one for each digit. The separation of contiguous digits does not work well with a neural network. It has to be performed as *preprocessing*.

The travelling salesman problem (100) This problem is representative for an entire collection of equivalent optimization problems. It can be solved with multilayer networks, but Hopfield networks (with feedback) perform better. A weight from neuron i to neuron j means that the path goes from city i to city j. If there are n cities, n^2 neurons are needed.

Graph partitioning (100) Another discrete optimization problem.

Noise removal from time series signals (100) The same as forecasting of time series, but the output neuron will now calculate a smoothed value of the time series instead of a future value.

Protein structure recognition (1 000) The input neurons will contain the results of various measurements of the protein, the output will identify which particular protein is found. Important in studies of the human genome.

Loan application scoring (1 000) The input neurons will represent various factors used in deciding about a loan. Income for example will be scaled between -1 for no income, and +1 for the maximum relevant income. Possession of a credit card will result in a value +1 or -1 for the corresponding input neuron. There will be one output neuron, deciding whether the loan is given or not. No loan will be -1, numbers between -1 and 1 can represent the certainty the program has about the eligibility for a loan. Similar techniques are used to recognize fraudulent tax return claims.

Image compression (1 000) Input is the original image, output the bits of the compressed image. This has to be done in real time, and neural networks offer an efficient and fast parallel architecture.

Medical diagnosis (1 000) The input neurons will represent symptoms, and their uncertainty or degree, e.g. fever can be very low, low, none, high or very high. Output will stand for possible illnesses. It is particularly salient in this case that neural networks can do fuzzy inference.

Sonar target recognition (10 000) Sonar images are very noisy. One pixel per neuron as input, and the output neuron can decide, e.g., whether a sea mine is present or not.

Navigation of a car (10 000) The input is an image of a line painted on the road surface. The output is the control of the steering wheel, gas, etc. This is a combination of pattern recognition and control.

Speech recognition (10 000) The inputs will be discretized spectral diagrams of phonemes. The output is one of the possible phonemes, one neuron per phoneme. The large numbers of inputs makes this a difficult problem. The separation of phonemes is usually done as preprocessing.

All these applications are in rapid development, and publications are fast becoming out of date. The reader who wants to know the most recent results should scan the Proceedings of a major neural network conference, for example the International Conference on Artificial Neural Networks, the IEEE International Conference on Neural Networks or the World Congress on Neural Networks.

2.7   Problems

1. For backpropagation, does the transfer function of the neurons have to be continuous with respect to the input to the neurons?

2. Does the backpropagation algorithm provide a rule for changing the learning rate η?

3. Is the weight matrix of a layered network symmetric?

4. How would you use a neural network in a control problem?

5. For a feedforward network, does an increase in the number of weights always result in a decrease of the error the network makes on a training set? Can you decrease the error by introducing more hidden neurons?

6. Are the weights in a network derived from a biological concept?

7. Financial applications use neural networks mainly for two tasks: time series forecasting and expert systems. Choose one of these two tasks and explain, for the chosen task, the decisions and choices that have to be made in progressing from the user requirements to the implementation in software of the neural network. How would you estimate the accuracy of the predictions by the network?

2.8   Project: Classificaton of Angiograms

Backpropagation is an algorithm with many tuning parameters. Examples are the learning rate and the number of hidden neurons. Because it is a basic gradient descent, function minimizing algorithm, any improvement on the gradient descent algorithm can be translated into a variant of the backpropagation algorithm.

Most commercial neural network packages offer some variants of the backpropagation algorithm. To choose and use such a package effectively, you need to know the field of non-linear function minimization quite well. Alternatively, you can try to program basic backpropagation yourself, observe its defects, and try to improve the algorithm.

This is what we try to do in this project. Some commercial backpropagation packages are very expensive, so if you program it yourself, you will have a valuable piece of software!

The example we choose for this project is the classification of angiograms with neural networks [89, 65]. Angiograms are x-ray images of human arteries. Our aim is to recognize automatically whether there is a blockage or a narrowing. This would be very useful in preventive medicine.

The angiograms are obtained by digitization of x-ray images. These digital images have to be preprocessed. This is an essential stage in all pattern classification. In preprocessing you employ all the knowledge you have about a problem in order to extract as much information as possible from the patterns to be classified. A typical example is an image where noise is present. This is the case for angiograms. You would use your knowledge about the x-ray machine that made the images, and about the digitization procedure, in order to estimate the noise that will be present. Then you will apply a filter to reduce this noise in the images [46]. We will assume that this has been done.

In a typical image you would see a tangle of arteries. It is possible to slide a window over the image, such that you see only a section of one or two arteries at a time. The size of the window depends on the scale of the image, and we will assume here that a 10 by 10 window is sufficient to isolate a section of just one artery. In figure 2.7 two typical situations are depicted. To the left is a section of artery as it would appear in a 10 by 10 window. To the right is a branching point between two arteries. These situations should be classified as negative (state of output neuron -1)

Figure 2.8 shows an example of a narrowing, and of a blockage. These cases should be classified as positive (state of output neuron +1).

In the examples of figures 2.7 and 2.8, the main artery crossed the window in the same direction every time. In general this will not happen, of course. We need to make a decision here whether the rotate the image as part of preprocessing, or to leave it as it is, and let the neural network take care of this. We choose the latter option here.

You need to create a number of prototypes now. Start by having the arteries run in the same direction, and draw six arteries with different widths, six prototypes of branches, and six prototypes of blockages. To

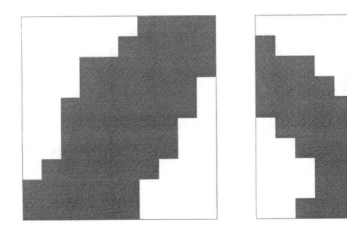

Figure 2.7: An example of a section of artery, and a branching point between two arteries.

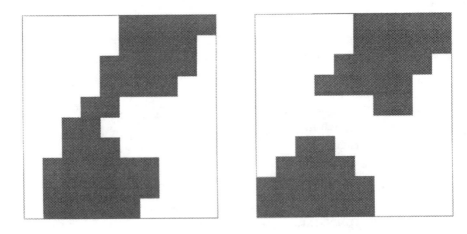

Figure 2.8: A narrowing and a blockage.

store the pictures, it is easiest if you have a bitmap editor. If not, you can just edit a 10 by 10 array of zeros and ones in a text editor. Your program will read and store the eighteen prototypes.

Now you have to choose a layered neural network architecture. It will have 100 input neurons, one for each pixel, and one output neuron, to indicate a positive or negative decision. The number of layers, and the number of hidden neurons in each layer, is there for you to explore. Program both a three and four layer network. For the three layer network, use either the formulas in section 2.3, or formulas (2.5) and (2.6). For the four layer network, you have to extend (2.5) and (2.6) with one more formula, for the update of the weights between the second and the first (input) layer.

How many hidden neurons should you choose? A proper answer would draw upon two fields, statistical estimation [17], and non-linear function approximation [88]. It is easier to experiment, however, and the following should give you some intuition. If there was only one neuron in each of the hidden layers, the output neuron would not contain more information than the neuron in the first hidden layer. This kind of network is called a perceptron, and has been well studied [85]. If you draw the prototypes as points in a 100-dimensional space, the perceptron can only classify them if the positive prototypes can be separated from the negative ones by a plane. This is unlikely to happen, so one neuron in each hidden layer is a bad idea.

Assume there was one hidden layer with 2^{100} hidden neurons, as many as there are possible images to classify. You could make a hidden neuron have state +1 for just one image, by a proper choice of the weights between the input and the hidden layer. A simple choice of the weights between hidden and output layer (which one?) then allows to classify each positive image with a +1 state of the output neuron, and the negative images with a -1 state. This network works perfectly, *but* you have to train all 2^{100} possible images!

The network with 2^{100} hidden neurons does not generalize. A network that does generalize will classify a pattern similar to a prototype that you have trained, in the same class that it was taught to classify the prototype. A network with few hidden neurons will show overgeneralization, it can only distinguish between a few classes. A network with too many hidden neurons will classify every prototype as a different class.

Try to find now the compromise that gives you sufficient generalization. In order to do this, divide your training prototypes in two. You use one half to train the network, and one half to test it. Use 20 training prototypes, and 20 to test. Fist find a good one hidden layer network, then try to find a two hidden layer network. Can you use less neurons in two hidden layers than in one hidden layer?

Fix the number of hidden neurons now. You may have observed, during training, that the changes to the weights Δw_{ij} become small as you approach the minimum of the error function (2.3). Try to counteract this by increasing the learning rate η.

An even more powerful way of speeding up convergence is the so called *momentum method*. In this method, you add to the change in weight Δw_{ij} a fraction α of the previous weight change. Do this for all weights. Start with $\alpha = 0.4$. It is not worth trying to find an optimal α, as this value would be dependent on the particular set of patterns that are trained, and on the number of hidden neurons.

You have now explored enough facets of the backpropagation algorithm to be able to use it in any practical pattern classification task. Do not forget that there exist many other pattern classification algorithms [35, 46].

Chapter 3

Neurons in the Brain

One of the main attractions of neural networks is that they offer a model of computation based on the processes taking place in the brain. This neurobiological computation is parallel, and may suggest an efficient parallel architecture for computing.

The idea of a computing brain has been proposed by McCulloch and Pitts [79]. In this chapter, we will only introduce sufficient neurobiological ideas to allow us to build a very simplified model of the brain. This model will be computationally rich enough for all applications of neural networks. For more detail, the reader can consult [39, 10, 84].

It is ironical that the neural network model furthest removed from reality, the multilayer network introduced in the preceding chapter, accounts for most of the applications of neural networks. This was our reason for introducing it so soon. The networks studied from this chapter onwards, are closer to neurobiological reality. The number of their applications is growing fast.

The multilayer networks from last chapter are well established, and the effort dedicated to them is more in development than in research. They represent the sure value. The networks studied in the main part of this book are in the growth area of neural networks. Among the many important applications is the modelling of time-dependent systems.

The neural network model derived in this chapter is a very general one. In the following chapters, this model will be simplified in several ways, each for a particular application field.

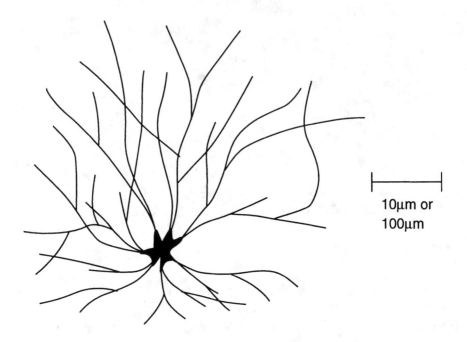

Figure 3.1: Sketch of a neuron. Dependent on the region of the brain and the function of the neuron, the diameter of the neuron is between a tenth and a hundredth of a millimeter.

3.1   The Physical Model

The brain is not a uniform mass. It has features, some of which are a centimeter large, some a millimeter, and others much smaller. We will not be concerned with large-scale features such as brain lobes. Our level of detail will be that of nerve cells. There exist even more low level descriptions of the brain, down to quantum mechanics [37, 38]. This is too detailed for our purpose.

A nerve cell is a well defined tiny region of the brain. The cells are separated by walls. The walls have a chemical composition different from the interior of the cell. A human brain consists of about 10^{11} nerve cells. A nerve cell is called a *neuron*.

Neurons have very irregular forms. They are far from spherical. This is illustrated in figure 3.1. The cell body (the central region in figure 3.1) is called the soma.

The protrusions of the soma are of two different kinds, called axons

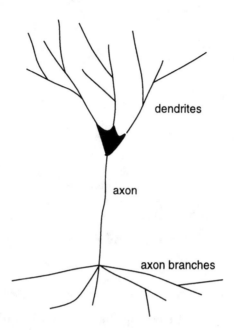

Figure 3.2: The neuron unfolded, to show the distinction between dendrites and axon branches.

and dendrites. The distinction is made clear if figure 3.1 is unfolded, as in figure 3.2.

How can axon branches be distinguished from dendrites? The axon branches end in what is called a synaptic bouton, see figure 3.3. The synaptic bouton will be used to transmit information from one neuron to another. We will sometimes use the term *synapse* instead of synaptic bouton.

Dendrites are surrounded by the synaptic boutons of other neurons, as in figure 3.4.

The neurons are highly interconnected, with each neuron having about 10^4 synaptic boutons connected to the dendrites of other neurons. Figure 3.5 shows how the neurons communicate with each other.

The number of neurons in the human brain, 10^{11}, is very large. It is of the same order of magnitude as the number of stars in a galaxy or the number of galaxies in the observable universe. Also, there are about 10^9 meters of axons, axon branches and dendrites. This is about 25 times the circumference of the earth. This shows that the brain is

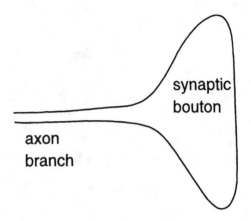

Figure 3.3: A synaptic bouton. Its size is about a tenth of a micron.

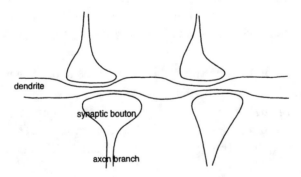

Figure 3.4: Synaptic boutons making contact with a dendrite

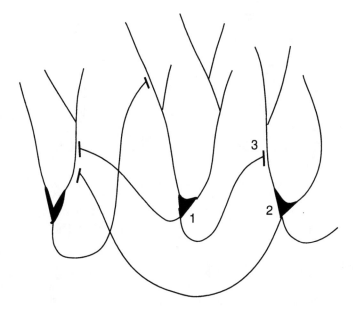

Figure 3.5: An example of connected neurons. Neuron 1 is connected to neuron 2 via synaptic bouton 3.

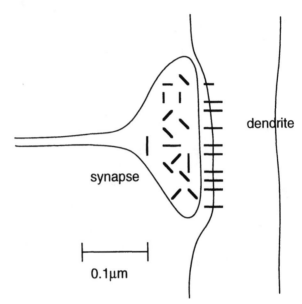

Figure 3.6: Transmitters and receptors in the synapse and dendrite.

a very complex system, and a true model of the brain would be very complicated. Building such a model is the task that neurobiologists face. Engineers, on the contrary, use simplified models that they can actually build.

Inside the neuron, in the soma, the axons, the synaptic boutons, and the dendrites is a fluid containing ions, mainly Na^-, K^+, and Cl^-. In the synapses are large molecules, called transmitters. In the dendritic wall are other large molecules, called receptors. This is shown in figure 3.6. Transmitters can pass the synapse wall because it acts as a membrane. They bind to receptors on the dendrite. This allows ions to flow from the synapse into the dendrite.

This describes how the neurons communicate through the exchange of ions. The ions carry electric charges. Because of the changing ion concentration inside the neurons, voltage spikes will travel in the neuron. In the next section, we will study these spikes.

3.2   Operation of the Neuron

Along the axon travels a spike or *action potential* of the form sketched in figure 3.7. The action potential is a travelling wave, and the figure

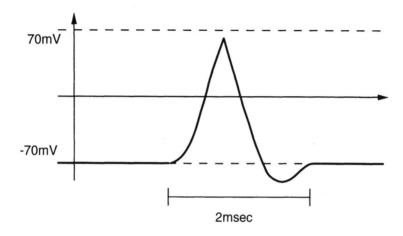

Figure 3.7: The shape of the travelling action potential at a particular moment in time.

gives the form of the wave, i.e. the spatial dimension is along the x-axis. At later moments in time, the waveform will be shifted.

The spike is not always present, and if there is no signal, the axon is at a resting potential of about -70 mV.

At a branch of the axon, the signal branches too, see figure 3.8. The amplitude remains the same. Remember, the axon is not a copper wire!

When the spike arrives at the synapse, neurotransmitters (complex molecules) are released. They bind to receptors and allow ions to flow into the dendrites [12]. The potential in the dendrite has the shape of figure 3.9 Depending on the neurotransmitters and the ions, these so-called *post-synaptic potentials* are excitatory or inhibitory, and can be more or less so, they have an efficacy. The excitatory postsynaptic potential is a result from the opening of an ion channel between synaptic bouton and dendrite that is permeable to Na^+ and K^+. Inhibitory postsynaptic potentials inhibit by preventing the membrane potential of the axon from reaching the threshold for spike generation. Inhibitory transmitters activate Cl^- or $K+$ channels [36].

The number of the postsynaptic potentials can be as large as the number of dendrites, about 10000. These post-synaptic potentials diffuse, or travel, through the dendrites towards the soma.

If there are more excitatory than inhibitory post-synaptic poten-

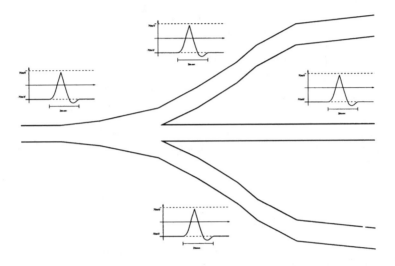

Figure 3.8: The action potential is distributed over the axon branches.

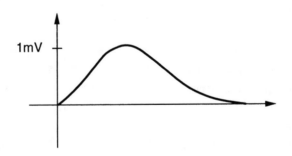

Figure 3.9: The potential in the dendrite.

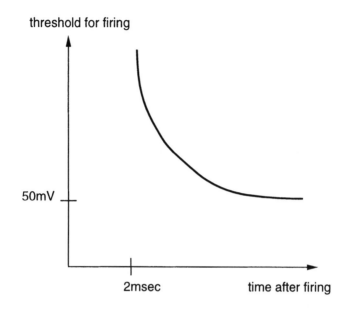

Figure 3.10: A time dependent threshold for firing.

tials, taking into account their efficacy, the soma can emit an action potential, and the process starts again as described at the start of this section.

The emission of the action potential depends on several factors. Before the neuron fires, a threshold may need to be transgressed. This threshold can be tens of millivolts, and can be time dependent. An example of time-dependency of the threshold is given in figure 3.10.

The firing of the neuron can also be stochastic, with noise involved. We will study this in chapter 8.

After emitting a spike, the neuron cannot emit another spike for 1 or 2 msec. This period is called the absolute refractory period. This implies that the maximum firing frequency of the neuron is between 500 and 1000 Hz. Experiments have shown that many neurons have firing frequencies around 50 Hz.

A neuron that has not fired, because it did not reach the threshold for firing, loses its potential to fire gradually. This is sometimes called leakage, in analogy with an electric current leaking away.

The description given here is approximate, and based on human neurons in the outer, wrinkled layer of the human brain. This layer,

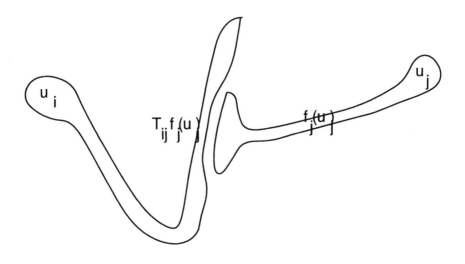

Figure 3.11: How neuron j influences neuron i.

called the cortex, is 2 to 4 mm thick.

3.3   The Mathematical Model

After simplifying neuronal activity to travelling voltage spikes, in the previous section, we can now build a mathematical model.

Remember that neuron j, as a cell, has a membrane. This membrane has a potential, denoted by u_j, see figure 3.11. On passing through the synapse, this ion concentration is altered to $T_{ij}f_j(u_j)$. The numbers T_{ij} can be positive or negative, as the ions can have a positive charge (Na^+), or a negative one (Cl^-).

The contributions to the potential in neuron i are summed over time, like ion concentrations can be summed. This sum is

$$\int_{-\infty}^{t} T_{ij}f_j(u_j)\,d\tau. \tag{3.1}$$

Remember that $u_j(t)$ is dependent on time. The expression 3.1 now has to be modified, because ions that arrive early depolarize, so that the contribution to u_i from u_j at time t is

$$\int_{-\infty}^{t} h(t-\tau)T_{ij}f_j(u_j)\,d\tau, \tag{3.2}$$

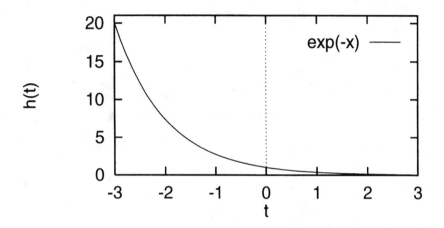

Figure 3.12: A possible decay function h(t).

with h a decay function equal to

$$h(t) = \frac{1}{\mu_i} e^{-\frac{t}{\mu_i}}.$$

The number μ_i is the time constant of the membrane, and $\mu_i > 0$. A large μ_i means that only the most recent contributions to the membrane potential are effective. As an be seen in figure 3.12, $h(t - \tau)$ is small for large negative τ, indicating that ions arriving early depolarize.

We will also assume that we can give direct, external inputs to the neurons, not coming from other neurons. For neuron i, this external input is denoted by I_i, see figure 3.13. The contribution of the external input to the potential u_i also fades away:

$$\int_{-\infty}^{t} h(t - \tau) I_i(\tau) \, d\tau.$$

Neuron i gets contributions to u_i from several other neurons, as in figure 3.14. These contributions are summed up, so that, for n neurons,

$$u_i(t) = \sum_{j=1}^{n} \int_{-\infty}^{t} h(t - \tau) T_{ij} f_j(u_j) \, d\tau + \int_{-\infty}^{t} h(t - \tau) I_i(\tau) \, d\tau. \quad (3.3)$$

The number u_i is all there is to know about neuron i, as far as we are concerned. It is the *state* of the neuron. Equation (3.3) is an

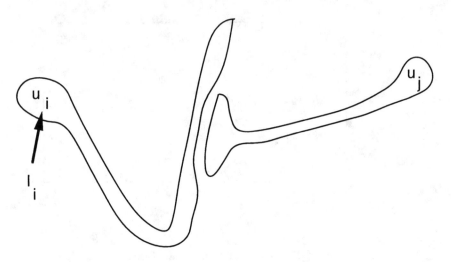

Figure 3.13: An external input I_i to neuron i.

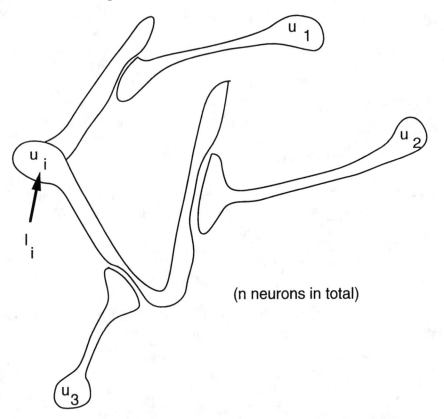

(n neurons in total)

Figure 3.14: Neuron i gets input from neurons 1, 2, and 3.

integral equation for the function u_i. Integral equations are difficult, therefore we will transform the equation into a differential equation, by taking the time derivative of the left- and right-hand sides of the equation. After some regrouping of terms, we obtain

$$\frac{du_i(t)}{dt} = \frac{d}{dt} \int_{-\infty}^{t} \frac{1}{\mu_i} e^{-\frac{t-\tau}{\mu_i}} \left[\sum_{j=1}^{n} T_{ij} f_j(u_j(\tau)) + I_i(\tau) \right] d\tau.$$

Using Leibnitz's rule for differentiation under the integral sign [97],

$$\frac{d}{dt} \int_{a(t)}^{b(t)} g(\tau, t) \, d\tau = \int_{a}^{b} \frac{\partial g}{\partial t} \, d\tau + g(b, t) \frac{db}{dt} - g(a, t) \frac{da}{dt}, \qquad (3.4)$$

we obtain

$$\frac{du_i(t)}{dt} = -\frac{1}{\mu_i} \int_{-\infty}^{t} \frac{1}{\mu_i} e^{-\frac{t-\tau}{\mu_i}} \left[\sum_{j=1}^{n} T_{ij} f_j(u_j(\tau)) + I_i(\tau) \right] d\tau \quad (3.5)$$

$$+ \frac{1}{\mu_i} \left[\sum_{j=1}^{n} T_{ij} f_j(u_j(t)) + I_i(t) \right] \qquad (3.6)$$

$$= -\frac{1}{\mu_i} u_i(t) + \frac{1}{\mu_i} \left[\sum_{j=1}^{n} T_{ij} f_j(u_j(t)) + I_i(t) \right]. \qquad (3.7)$$

This equation for the state u_i of neuron i can be written down for all n neurons, so that we obtain a system of differential equations

$$\mu_i \frac{du_i(t)}{dt} = -u_i(t) + \sum_{j=1}^{n} T_{ij} f_j(u_j(t)) + I_i(t),$$

$$i = 1, \ldots, n, \quad \mu_i > 0. \qquad (3.8)$$

This is the fundamental system of differential equations that governs the operation of the neural network. We will study it in detail in the next chapter.

3.4 The Transfer Function

The function f_j introduced in (3.1) is called the transfer function. It models what happens in the synapses, and it has been the subject of

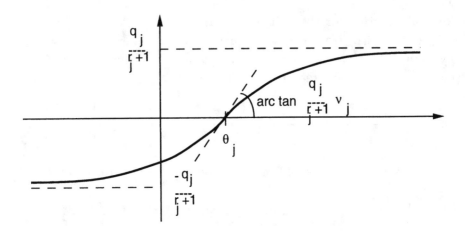

Figure 3.15: The transfer function f_j.

many neurobiological studies [58, 84]. The function can be approximated as

$$f_j(u_j) = \frac{q_j}{r_j + 1}\tanh(\nu_j(u_j - \theta_j)). \tag{3.9}$$

A sketch of this function can be seen in figure 3.15. The parameter q_j is the charge delivered per impulse, or spike, r_j is the refractory period, θ_j is a threshold, and ν_j determines the slope. This gives for the dimension of f_j a charge per second, or the dimension of a current. The function has an asymptote at $\frac{q_j}{r_j+1}$ for $u_j \to \infty$, and at $-\frac{q_j}{r_j+1}$ for $u_j \to -\infty$. The intercept with the horizontal axis is at θ_j, and the angle of the curve with this axis at this point is arc $\tanh\frac{q_j\nu_j}{r_j+1}$. The threshold θ_j is often subject to noise, but we will not discuss this here.

It is not strictly necessary to use the tanh function, $\frac{2}{\pi}$arctan works well too, for example.

The function tanh is not always readily available, certainly not in a VLSI implementation of a neural network. The following approximation can be used

$$\tanh x = x - \frac{x^3}{3} + \frac{2}{15}x^5 - \frac{17}{315}x^7 + \dots, \quad |x| < \frac{\pi}{2}. \tag{3.10}$$

If $|x| > \pi/2$, the following formula may have to be used repeatedly, before (3.10) can be used

$$\tanh x = \frac{2\tanh\frac{x}{2}}{1 + \tanh^2\frac{x}{2}}. \tag{3.11}$$

The occurrence of the function tanh in models of neural networks in the brain, makes these models nonlinear systems. The negative side of this is that they become much more difficult to analyze than linear systems, the positive side is that many phenomena occur that are unknown in linear systems. This makes the brain such a rich and interesting system to study, even if we are restricted to the simplified model (3.8).

3.5 Problems

1. Are neural networks closely based on brain models better than networks loosely based on such biological models? Discuss this from the standpoint of an engineer faced with a problem that can be solved by neural networks.

2. Has the function sgn in

$$x_i(t+1) = \text{sgn}\left(\sum_{j=1}^{n} T_{ij}x_j(t)\right), \quad i = 1,\ldots,n$$

any biological relevance?

3. Describe, in a non-mathematical way, what biological principles are used in the derivation of an analog neural network model. Mention for every biological factor the corresponding neural network parameter.

4. Are the weights in a network derived from a biological concept?

5. Is the absolute refractory period a bound on the time interval in which a neuron can switch from a passive to an active state?

3.6 Project: Neural Oscillators

Models of part of the brain have to take into account the *topology* of the connections, because not every neuron is connected to every other neuron. In this project, we will study a little network of neurons connected in a line, where each neuron has another neuron associated with it [102]. We will call the neurons in line the u-neurons, and denote

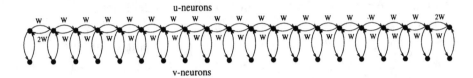

Figure 3.16: A network of 20 u-neurons in line, each with their associated v-neuron. The symbols on the arrows are the weights between the u-neurons.

their state by $u_i(t), i = 1, \ldots, 20$. The associated neurons are called v-neurons, and their states are $v_i(t), i = 1, \ldots, 20$, see figure 3.16.

We can now write down two systems of differential equations similar to (3.8), one for each type of neuron,

$$\frac{du_i}{dt} = -u_i + \tanh(u_i - \beta v_i) + \sum_{j=1}^{20} T_{ij} \tanh(u_j) + I_i,$$

$$\mu \frac{dv_i}{dt} = -v_i + \tanh(\alpha u_i),$$

$$i = 1, \ldots, 20. \tag{3.12}$$

The simplified transfer function tanh has been used. Only the u-neurons can get external inputs I_i. The coupling between the u and v neurons is modeled by the numbers α and β. For appropriate values of these constants, the v-neurons will cause oscillations in the states of the u-neurons. The period of these oscillations will be influenced by the time constant μ.

The weights T_{ij} between adjacent neurons all have the value W, except at the end of the chain. Formally, this can be expressed as

$$T_{ij} = \begin{cases} W, & j = i+1, j = i-1, 1 < i < 20, \\ 2W, & j = 2, i = 1, \\ 2W, & j = 19, i = 20, \\ 0, & \text{elsewhere.} \end{cases} \tag{3.13}$$

This is a very crude model of part of the cortex [102].

The equations (3.12) have to be solved by numerical approximation of the time derivative. This is explained in detail in section 4.2, but here we will use the simple formula

$$u_i(t+h) = u_i(t) + h[-u_i(t) + \tanh(u_i(t) - \beta v_i(t))$$

$$+ \sum_{j=1}^{20} T_{ij} \tanh(u_j(t)) + I_i],$$

$$v_i(t + h) = v_i(t) + \frac{h}{\mu}[-v_i(t) + \tanh(\alpha u_i(t))],$$

$$i = 1, \ldots, 20. \tag{3.14}$$

These formulae allow the calculation of $\mathbf{u}(t+h)$ and $\mathbf{v}(t+h)$ if $\mathbf{u}(t)$ and $\mathbf{v}(t)$ are known, for small time increments h. The states of all neurons have to be known at some initial time t_0.

To understand the interplay between μ, α, and β, simulate the following system with only one u-neuron and one v-neuron and no external input,

$$u(t + h) = u(t) + h[-u(t) + \tanh(u(t) - \beta v(t))],$$

$$v(t + h) = v(t) + \frac{h}{\mu}[-v(t) + \tanh(\alpha u(t))]. \tag{3.15}$$

Choose arbitrary starting values for $u(0)$ and $v(0)$, and choose $h = 0.01$. Plot the trajectory of the point $(u(ih), v(ih)), i = 0, \ldots, 1000$ in the plane. Vary μ, α, and β, and see how this trajectory changes. Find values for these parameters such that an oscillation is sustained. Estimate the period of this oscillation.

Now simulate the network with 20 u-neurons and 20 v-neurons. Use the values for μ, α, and β that you found in the previous simulations. Assume that all the inputs I_i are zero. Choose a value for W, so that the weights in (3.13) are determined. Use random values uniformly between +1 and -1 for $\mathbf{u}(0)$ and $\mathbf{v}(0)$. Plot the 20 waveforms $\mathbf{u}(ih), i = 0, \ldots, 1000$. Do the waveforms synchronize? Is there a phase lock or a phase shift in time? You may need to simulate for a shorter or a longer time, dependent on the frequency of the oscillations. If you do not observe synchrony, you may need to change the value of W, or make the I_i non-zero.

Some weights on the end of the chain of neurons were doubled from W to $2W$. Experiment with this, changing $2W$ to W, or making both weights between the two u-neurons at the ends equal to $2W$. Do the neurons still synchronize?

What we did up to now was to fine tune the system, setting up the parameters so that the neurons oscillate in synchrony or quasi syn-

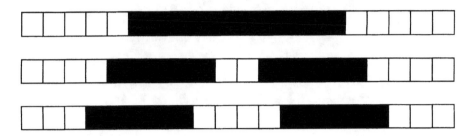

Figure 3.17: Three input configurations for the 20 u-neurons in the network of figure 3.16. The black squares represent non-zero input.

chrony. The actual aim of the project is to change the external inputs I_i for some neurons, and observe whether synchrony is maintained.

Start by having $I_i = 0$ for the five u-neurons at both ends of the chain, and $I_i = 1.0$ for the ten neurons in the middle. Plot the 20 waveforms of the u-neurons. Is there synchrony? You may have to change the value of the I_i of the middle neurons a bit. How long does it take the neurons to synchronize?

Is is possible to plot the correlation between neurons i and j, by calculating $u_i(t)u_j(t)$ as time evolves? This plot may be more informative than the waveforms themselves if there are slow phase changes between u_i and u_j.

The non-zero input for the ten neurons in the middle can be visually represented as in figure 3.17. Now try the other two input configurations from figure 3.17. Verify whether there is synchrony between neurons in the two different blocks of inputs, for example by plotting $u_7(t)u_{14}(t)$ for the middle input configuration, and $u_6(t)u_{15}(t)$ for the bottom input configuration.

In general, how far can the two black bars be brought apart without losing the synchrony between their u-neurons? You may want to try a two-dimensional version of this, where you have two shapes in a two-dimensional grid, and you vary the distance between them.

As a last experiment, alter the slope of the tanh function in (3.14) by using $\tanh(\nu x)$ instead of $\tanh(x)$, as explained in section 3.4. How are the dynamics of the network altered?

In this project, you have done a simple simulation with a crude model of part of the cortex. Such simulations form the basis for computational models of several regions in the brain.

Chapter 4

The Fundamental System of Differential Equations

In this chapter, we will study the system of differential equations that was established as a model of a neural network in last chapter. It is the model that is most true to life, or neurobiological reality. In later chapters, we will encounter more simplified models.

Because the system of differential equations is a sophisticated model, it will show the most rich and interesting behaviour of all neural network models. This behaviour consists of the evolution in time of the state of the neurons. It is called *dynamic* behaviour, because it changes in time. The system of first order differential equations that describes this change in time is called a *dynamical system*. This chapter is about the particular dynamical system that governs the evolution in time of a neural network.

The nonlinear transfer function of the neuron makes the dynamical system nonlinear. The system can be linearized in a small region of state space, but we will not use this approximation, because we are interested in the *global* behaviour of the network. We want a comprehensive view of the state space, and knowledge about what states are typical for the network. These global properties will be derived from local knowledge about the synapses.

Our main tool will be functions similar to the energy function in physics. As the neural network evolves, the energy function will be minimized. This replaces the evolution of the network in a high dimensional state space with the minimization of a scalar energy function, some-

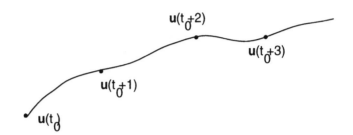

Figure 4.1: The unique solution of the dynamical system , determined by the initial condition at time t_0.

thing that is much easier to visualize. It also allows, by inverting this procedure, to construct a neural network that minimizes a particular function. This is an important application area for neural networks.

4.1   Existence of a Solution

In the previous chapter, we have derived the following model for the neural network:

$$\mu_i \frac{du_i(t)}{dt} = -u_i(t) + \sum_{j=1}^{n} T_{ij} f_j(u_j(t)) + I_i(t), \quad i = 1, \ldots, n, \quad \mu_i > 0.$$

$$(4.1)$$

We will not assume automatically that f_j is tanh anymore, because the function f_j will usually be implemented by some electronic circuit. This circuit may for example implement a piecewise linear approximation to tanh.

The system (4.1) is of the form

$$\dot{\mathbf{u}} = g(\mathbf{u}, t), \quad \mathbf{u} \in R^n,$$

$$(4.2)$$

the equation for a general dynamical system. From the theory of dynamical systems [2, 13, 14, 101], we can write down two important theorems about the existence of a solution to (4.2).

Theorem 4.1 *For sufficiently small $|t - t_0|$, (4.2) has a unique solution that verifies the initial condition $\mathbf{u}(t_0) = \mathbf{u}_0 \in R^n$.*

This theorem implies that, given an initial condition, there is only one way that the neural network can evolve in time. This is important for the numerical approximation of the solution to (4.2).

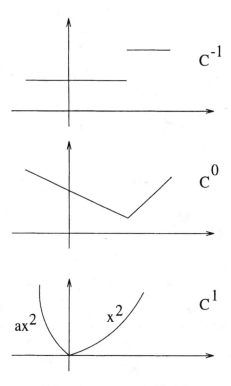

Figure 4.2: Functions differing in smoothness from C^{-1} to C^1.

The condition that $|t - t_0|$ be sufficiently small guarantees the invertibility of the right hand side of (4.2) and (4.1).

The right hand side of (4.1) depends on several parameters: the weights T_{ij}, and through f_j it depends on the parameters $q_j, r_j, \nu_j, \theta_j$. This dependence is called C^r smooth if the right hand side, after r differentiations, is continuous. If the function is not differentiable, it is called C^{-1} smooth. In figure 4.2 are some examples of functions of varying smoothness.

The following theorem establishes how smooth the time evolution of the neural network is, given the smoothness of the right hand side of (4.1).

Theorem 4.2 *If the right hand side of (4.1) is C^r with respect to the parameters t and $\mathbf{u}$, then (4.1) has a solution obeying the initial condition $\mathbf{u}(t_0) = \mathbf{u}_0$, and C^r with respect to t_0, $\mathbf{u}_0$, the parameters of f_j, and t. The solution is C^{r-1} with respect to $\mathbf{u}$. For this theorem to*

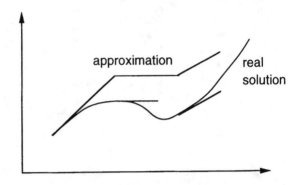

Figure 4.3: Reconstruction of a function from a known derivative.

hold, $|\mathbf{u} - \mathbf{u}_0|$ and $|t - t_0|$ have to be sufficiently small.

The importance of this theorem is again in engineering applications. When the functions f_j are approximated by a piecewise linear function, for example, the theorem tells how smooth the network will evolve in state space.

4.2 Numerical Solution of the Equations

In order to follow the evolution in time of a neural network, it is usually necessary to solve the system (4.1) numerically. We will use a Runge-Kutta method [4, 62]. It works well if the functions f_j are sufficiently smooth.

The principle is as follows. Starting from an initial guess, the solution to (4.1) is constructed using the known value of the derivative, namely the value of the right hand side. This is illustrated in figure 4.3.

For clarity, we will now change notation. If there is only one neuron, the system (4.1) contains only one equation of the form $\dot{x} = f(x,t)$. Assume that x_0 is the initial condition, at time $t = t_0$. The function $x(t)$ will then be approximated by $x_1 = x(t_0 + h), x_2 = x(t_0 + 2h)$, etc. The increment h is the stepsize. Refer to figure 4.4. If we now denote $x(t_0 + ih) = x(t_i) = x_i$, the Runge Kutta formula is

$$x_{i+1} = x_i + \frac{1}{2}(k_1 + k_2) + \mathcal{O}(h^3), \quad i = 0, 1, \ldots,$$
$$k_1 = hf(x_i, t_i),$$

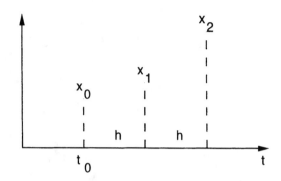

Figure 4.4: The approximation of $x(t)$ at discrete times $t_0 + h, t_0 + 2h, \ldots$

$$k_2 = hf(x_i + k_1, t_i + h). \tag{4.3}$$

The term $\mathcal{O}(h^3)$ gives an estimate of the error. It means that, for small h, the error is $a_3h^3 + a_4h^4 + a_5h^5 + \ldots$, with the a_i constants independent of h, but dependent on f. For an alternative definition, see page 13.

For 2 neurons, there are two differential equations

$$\dot{x} = f(x, y, t),$$
$$\dot{y} = g(x, y, t). \tag{4.4}$$

The Runge Kutta formula is now

$$
\begin{aligned}
x_{i+1} &= x_i + \frac{1}{2}(k_1 + k_2) + \mathcal{O}(h^3), \quad i = 0, 1, \ldots, \\
y_{i+1} &= y_i + \frac{1}{2}(l_1 + l_2) + \mathcal{O}(h^3), \quad i = 0, 1, \ldots, \\
k_1 &= hf(x_i, y_i, t_i), \\
l_1 &= hg(x_i, y_i, t_i), \\
k_2 &= hf(x_i + k_1, y_i + l_1, t_i + h), \\
l_2 &= hg(x_i + k_1, y_i + l_1, t_i + h). \tag{4.5}
\end{aligned}
$$

For more neurons, meaning more differential equations, it is easier to generalize these formulas yourself than to write down the general equations.

We mention now, for 2 neurons, a formula with a smaller error, $\mathcal{O}(h^5)$.

$$\dot{x} = f(x, y, t),$$

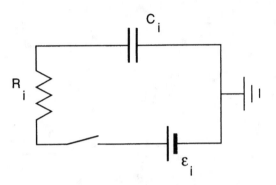

Figure 4.5: A simple RC circuit.

$$
\begin{aligned}
\dot{y} &= g(x, y, t), \\
x_{i+1} &= x_i + \frac{1}{6}(k_1 + 2k_2 + 2k_3 + k_4) + \mathcal{O}(h^5), \quad i = 0, 1, \ldots, \\
y_{i+1} &= y_i + \frac{1}{6}(l_1 + 2l_2 + 2l_3 + l_4) + \mathcal{O}(h^5), \quad i = 0, 1, \ldots, \\
k_1 &= hf(x_i, y_i, t_i), \\
l_1 &= hg(x_i, y_i, t_i), \\
k_2 &= hf(x_i + k_1/2, y_i + l_1/2, t_i + h/2), \\
l_2 &= hg(x_i + k_1/2, y_i + l_1/2, t_i + h/2), \\
k_3 &= hf(x_i + k_2/2, y_i + l_2/2, t_i + h/2), \\
l_3 &= hg(x_i + k_2/2, y_i + l_2/2, t_i + h/2), \\
k_4 &= hf(x_i + k_3, y_i + l_3, t_i + h), \\
l_4 &= hg(x_i + k_3, y_i + l_3, t_i + h).
\end{aligned}
\tag{4.6}
$$

Runge Kutta methods can be a bad choice, for example if the solution is oscillating in time. In this case, other methods have to be used, e.g. Gear methods. Any good software library should offer Runge Kutta and Gear methods for solving systems of ordinary differential equations.

4.3 An Analog Circuit

In figure 4.5, an RC Circuit is drawn. The equation for the potential

V_i is

$$\epsilon_i = R_i C_i \frac{dV_i}{dt} + V_i.$$

If we put, for the time constant,

$$R_i C_i = \mu_i, \tag{4.7}$$

the equation becomes

$$\mu_i \frac{dV_i}{dt} = -V_i + \epsilon_i.$$

Consider now again the fundamental system of differential equations (4.1),

$$\mu_i \frac{du_i(t)}{dt} = -u_i(t) + \sum_{j=1}^{n} T_{ij} f_j(u_j(t)) + I_i(t), \quad i = 1, \ldots, n, \quad \mu_i > 0.$$

$$\tag{4.8}$$

With the convention (4.7) for the time constant, this becomes

$$C_i \frac{du_i}{dt} = -\frac{u_i}{R_i} + \sum_{j=1}^{n} \frac{T_{ij}}{R_i} f_j(u_j) + \frac{I_i}{R_i}. \tag{4.9}$$

The left hand side of this equation can be interpreted as an input current charging a capacitor C_i to a potential u_i. In the right hand side, $-u_i/R_i$ is a leakage current, and I_i/R_i an input from outside the network. If the sum term stands for input currents from other neurons, $f_j(u_j)$ is a potential, the output of an amplifier, and T_{ij}/R_i are conductances. The numbers R_i are just scaling factors here, not resistances.

Conductances are always positive, so a solution has to be found for negative synapses. This is done by inverting the output of the amplifier, giving a signal $f_j(u_j)$ as well as $-f_j(u_j)$, see figure 4.6.

This analysis leads to the analog electric network sketched in figure 4.7. It is the basic layout of many chip implementations of neural networks, even digital implementations, [48, 40, 81, 22]. Remark that all the synapses take up much more space than all the neurons.

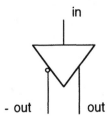

Figure 4.6: Negative synapses implemented using an inverter.

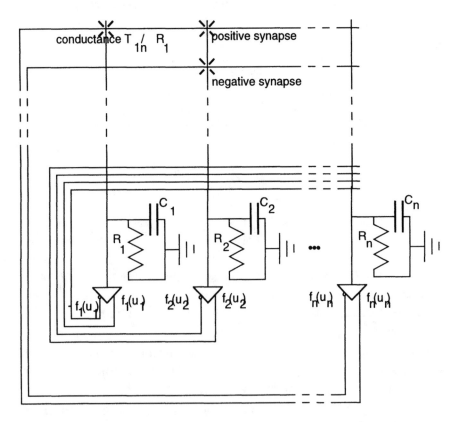

Figure 4.7: A basic electronic circuit for a neural network.

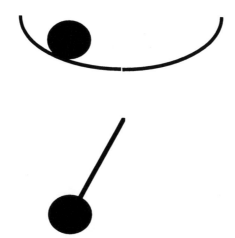

Figure 4.8: Two physical systems that evolve towards an energy minimum.

4.4 The Energy Function

The time evolution, or dynamic behaviour, of a mechanical system is governed by a differential equation, Newtons law,

$$\ddot{\mathbf{x}} = \frac{F}{m},$$

relating the second time derivative of the place coordinates $\mathbf{x}$ to the forces F and the mass m.

When the system evolves in time, it evolves towards an energy minimum. This is illustrated in figure 4.8 for two mechanical systems, a ball rolling down a bowl, and a pendulum.

A mechanical system where the outside forces are independent of time is called *autonomous*. In this section we will assume that the neural network is autonomous, i.e.

$$\frac{dI_i}{dt} = 0,$$

or the external inputs I_i are independent of time.

There exists no recipe for finding the energy function of a neural network. Moreover, the dynamic equations (4.1) for a neural network

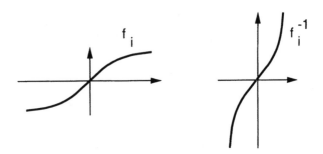

Figure 4.9: The transfer function f_i and its inverse.

contain first order derivatives in time, whereas Newtons law contains second order time derivatives. We will therefore propose an expression for the energy function, and verify whether it decreases as the system evolves in time. The proposal is [50]

$$E = -\frac{1}{2}\sum_{i=1}^{n}\sum_{j=1}^{n}T_{ij}f_i(u_i)f_j(u_j) - \sum_{i=1}^{n}I_if_i(u_i) + \sum_{i=1}^{n}\int_{x=0}^{x=f_i(u_i)}f_i^{-1}(x)\,dx.$$

(4.10)

If the transfer functions were linear, the first term would be of second order in the state variables u_i. The second term, too, is usual for an external force I_i.

It is the third term that is rather surprising. The symbol f_i^{-1} is the inverse function of f_i, defined by $f_i^{-1}(f_i(x)) = x$, see also figure 4.9. Consider the upper limit of the integral

$$x = f_i(u_i) = \frac{q_i}{r_i+1}\tanh(\nu_i(u_i-\theta_i)),$$

or

$$\nu_i(u_i-\theta_i) = \operatorname{arctanh}\frac{r_i+1}{q_i}x,$$

hence

$$u_i = \theta_i + \frac{1}{\nu_i}\operatorname{arctanh}\frac{r_i+1}{q_i}x = f_i^{-1}(x).$$

Using this expression, we have for the integral in the energy function, with formula 2.741.3 from [47],

$$\int_0^x f_i^{-1}(x)\,dx = \int_0^x \theta_i\,dx + \frac{1}{\nu_i}\int_0^x \operatorname{arctanh}\frac{r_i+1}{q_i}x\,dx$$

$$\begin{aligned}
&= \theta_i x + \frac{1}{\nu_i}\left\{ x\,\mathrm{arctanh}\frac{r_i+1}{q_i}x \right.\\
&\quad + \frac{q_i}{2(r_i+1)}\ln\left[\left(\frac{q_i}{r_i+1}\right)^2 - x^2\right]\\
&\quad \left. - \frac{q_i}{2(r_i+1)}\ln\left[\left(\frac{q_i}{r_i+1}\right)^2\right]\right\}\\
&= \theta_i x + \frac{1}{\nu_i}\left\{ x\,\mathrm{arctanh}\frac{r_i+1}{q_i}x \right.\\
&\quad \left. + \frac{q_i}{2(r_i+1)}\ln\left[1 - \left(\frac{r_i+1}{q_i}x\right)^2\right]\right\}. \quad (4.11)
\end{aligned}$$

Remark that

$$\mathrm{arctanh}\,y = \frac{1}{2}\ln\frac{1+y}{1-y}.$$

The result of this calculation is that the energy function (4.10) is just a complicated expression in $f_i(u_i), i = 1,\ldots,n$, with nothing indeterminate.

Let us now abbreviate $v_i = f_i(u_i), i = 1,\ldots,n$, so that 4.10 becomes

$$E = -\frac{1}{2}\sum_{i,j}T_{ij}v_iv_j - \sum_i I_iv_i + \sum_i \int_0^{v_i} f_i^{-1}(x)\,dx. \quad (4.12)$$

Theorem 4.3 *If the matrix T is symmetric, has zero diagonal, and if the functions f_j are monotonically increasing, then the function*

$$E = -\frac{1}{2}\sum_{i,j}T_{ij}v_iv_j - \sum_i I_iv_i + \sum_i \int_0^{v_i} f_i^{-1}(x)\,dx$$

is an energy function.

Proof. In expression 4.12, v_i is a function of time, and dE/dt will be the evolution of energy with time. Using the chain rule,

$$\begin{aligned}
\frac{dE}{dt} &= \sum_{k=1}^n \frac{\partial E}{\partial v_k}\frac{dv_k}{dt}\\
&= \sum_{k=1}^n \frac{\partial}{\partial v_k}\left(-\frac{1}{2}\sum_{i=1}^n\sum_{j=1}^n T_{ij}v_iv_j - \sum_{i=1}^n I_iv_i + \sum_{i=1}^n \int_0^{v_i} f_i^{-1}(x)dx\right)\frac{dv_k}{dt}
\end{aligned}$$

$$
= \sum_{k=1}^{n} \left[\frac{\partial}{\partial v_k} \left(-\frac{1}{2} \sum_{i \neq k}^{n} T_{ik} v_i v_k - \frac{1}{2} \sum_{j \neq k}^{n} T_{kj} v_k v_j - \frac{1}{2} T_{kk} v_k^2 \right) \right.
$$

$$
\left. - I_k + \frac{\partial}{\partial v_k} \int_0^{v_k} f_k^{-1}(x) \, dx \right] \frac{dv_k}{dt}
$$

$$
= \sum_{k=1}^{n} \left[\left(-\frac{1}{2} \sum_{i \neq k}^{n} T_{ik} v_i - \frac{1}{2} \sum_{j \neq k}^{n} T_{kj} v_j - T_{kk} v_k \right) \right.
$$

$$
\left. - I_k + f_k^{-1}(v_k) \right] \frac{dv_k}{dt}, \tag{4.13}
$$

where we have also used Leibnitz's rule for differentiation under the integral sign [97].

We will assume now that the matrix T is symmetric, $T_{ij} = T_{ji}, i = 1, \ldots, n, j = 1, \ldots, n$, and has zero diagonal, $T_{ii} = 0, i = 1, \ldots, n$. In biological terms, these assumptions mean that the synapse from neuron i to j is the same as the synapse from j to i, and that no axon branches connect to the neurons own dendrites. This is not very plausible biologically, but helps a great deal in the engineering design of neural networks. The mathematical analysis is much simpler when T is symmetric and has zero diagonal. Asymmetric T and non-zero diagonal T matrices are a topic of current research. We will relax the conditions on T in Chapter 6 and 8.

Continuing the derivation of dE/dt, and using (4.1), we find

$$
\frac{dE}{dt} = \sum_{k=1}^{n} \left(-\sum_{i=1}^{n} T_{ki} v_i - I_k + f_k^{-1}(v_k) \right) \frac{dv_k}{dt}
$$

$$
= \sum_{k=1}^{n} \left(-\sum_{i=1}^{n} T_{ki} f_i(u_i) - I_k + f_k^{-1}(f_k(u_k)) \right) \frac{dv_k}{dt}
$$

$$
= -\sum_{k=1}^{n} \mu_k \frac{du_k}{dt} \frac{dv_k}{dt}
$$

$$
= -\sum_{k=1}^{n} \mu_k \frac{du_k}{dt} \frac{df_k(u_k)}{dt}
$$

$$
= -\sum_{k=1}^{n} \mu_k \frac{df_k}{du_k} \left(\frac{du_k}{dt} \right)^2. \tag{4.14}
$$

The time constants $\mu_i > 0, k = 1, \ldots, n$, and if the transfer functions are monotonically increasing, $df_k/du_k > 0$. Moreover, $(du_k/dt)^2 \geq 0$,

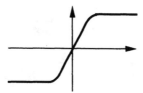

Figure 4.10: A function with zero derivative for large absolute values of its input.

and thus $dE/dt < 0$, showing that the energy decreases in time. During the operation of the network, energy will decrease until it does not change any more.
Q.E.D.

From equation (4.14), we can also deduce

$$\frac{dE}{dt} = 0 \Rightarrow \frac{du_i}{dt} = 0, \quad i = 1, \ldots, n. \tag{4.15}$$

This means that stationary points for the energy are stationary points for the *whole* neural network.

For some approximations of the transfer function, $df_k/dt = 0$ in some interval, as illustrated in figure 4.10. Even in this case, the energy decreases during the operation of the network.

4.5 Stability of Equilibria

When the network is observed to be in a particular initial state $\mathbf{u}_0$ at time t_0, the numbers $u_1(t_0), u_2(t_0), \ldots, u_n(t_0)$ all have values. The vector containing those values is called the state vector of the neural network. If there are n neurons, this is a vector in an n-dimensional space, the state space. In figure 4.11, you can see the trajectory in the state space, between time t_0 and t_2. For many applications of neural networks, we want to know what happens to the trajectory as time proceeds, $t \to \infty$. The most useful situation, termed convergence in engineering terms, is sketched in figure 4.12. Other possibilities, often a nightmare for the engineer, are in figure 4.13.

Analogous to a ball rolling in a bowl, the equilibrium can be *stable* or *unstable*, see figure 4.14. Different initial conditions, close to a stable equilibrium, will tend to make the system evolve towards this equilibrium. An unstable equilibrium will only be reached starting from one

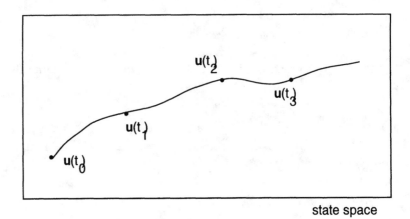

state space

Figure 4.11: Time evolution in state space.

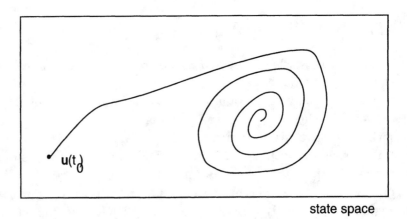

state space

Figure 4.12: Convergence.

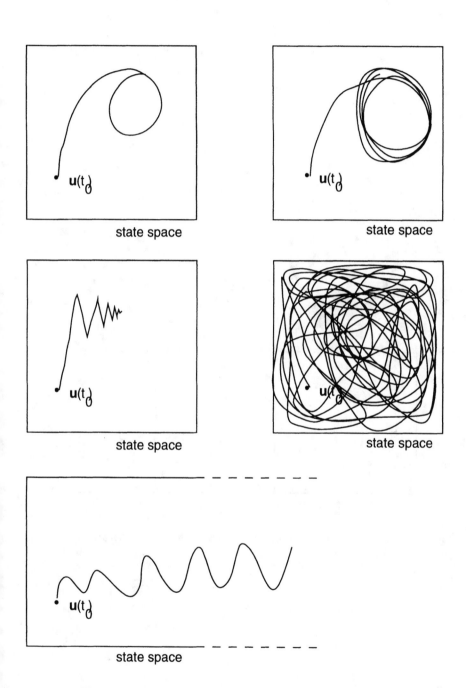

Figure 4.13: Some trajectories in state space as $t \to \infty$.

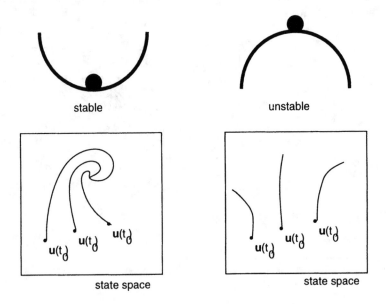

Figure 4.14: Stable and unstable equilibria. The trajectories are for different initial conditions.

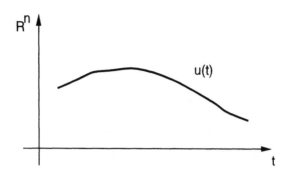

Figure 4.15: Time evolution of the state of one neuron.

particular initial condition, and then only if there is no noise. This is possible in a digital system when the variables are discrete. In the presence of noise, the system will never be in an unstable equilibrium.

A word of caution is necessary here. The dynamical systems we are studying are high dimensional if there are many neurons. The mathematical properties of the state spaces of these systems are very complicated, and are often open research problems. For example, next to stable and unstable equilibria, saddle points play an important role. In this book we only discuss the very basics that seem necessary to us for applying and understanding neural networks. For more detail, the reader should consult [2, 13, 14, 73, 83, 30, 57].

Time can be added as an extra dimension in the state space. This is illustrated in figure 4.15

We will assume for simplicity from now on that there is no external input, and that all transfer functions go through the origin. This means $I_i(t) = 0, t \geq t_0 \geq 0$ and $\theta_i = 0$. The fundamental system of differential equations is now

$$\mu_i \frac{du_i(t)}{dt} = -u_i(t) + \sum_{j=1}^{n} T_{ij} f_j(u_j(t)), \quad i = 1, \ldots, n, \quad \mu_i > 0. \quad (4.16)$$

The trajectory $u_i(t) = 0, i = 1, \ldots, n, t \geq t_0 \geq 0$ is a solution of this system. This solution is independent of time, and is an equilibrium of (4.16).

More formally, we define an equilibrium as a point in state space for which

$$\frac{du_i}{dt} = 0, \quad i = 1, \ldots, n. \quad (4.17)$$

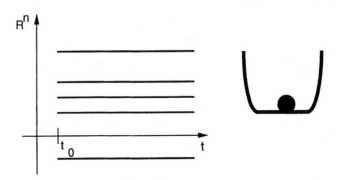

Figure 4.16: A stable equilibrium.

As we will investigate convergence, we need a notion of distance in state space. For two vectors $\mathbf{u}^1$ and $\mathbf{u}^2$, the Euclidean distance is

$$\sqrt{(u_1^1 - u_1^2)^2 + (u_2^1 - u_2^2)^2 + \ldots + (u_n^1 - u_n^2)^2}. \qquad (4.18)$$

Another distance is

$$|u_1^1 - u_1^2| + |u_2^1 - u_2^2| + \ldots + |u_n^1 - u_n^2|. \qquad (4.19)$$

In general, we will denote by $|\mathbf{u}^1 - \mathbf{u}^2|$ any function obeying

$$\begin{aligned}
|\mathbf{u}^1 - \mathbf{u}^2| &= 0 \Leftrightarrow \mathbf{u}^1 = \mathbf{u}^2, \\
|\mathbf{u}^1 - \mathbf{u}^2| &= |\mathbf{u}^2 - \mathbf{u}^1|, \\
|\mathbf{u}^1 - \mathbf{u}^3| &\leq |\mathbf{u}^1 - \mathbf{u}^2| + |\mathbf{u}^2 - \mathbf{u}^3|.
\end{aligned} \qquad (4.20)$$

The symbol $\mathbf{o}$ is the null vector or origin in n dimensions. By the norm of a vector, we mean the distance to the origin, notation $|\mathbf{u}|$.

Here are the three definitions of stability we will use. They are written down for the equilibrium $\mathbf{o}$ of (4.16). The trajectory staying in the origin for any $t \geq t_0 \geq 0$ is denoted by $\mathbf{o}(t)$.

The origin $\mathbf{o}(t)$ is a *stable* equilibrium if it is possible to force solutions $\mathbf{u}(t, t_0, \mathbf{u}^0)$ of (4.16) to remain as closely as desired to the equilibrium for all $t \geq t_0 \geq 0$ by choosing $\mathbf{u}^0$ sufficiently close to $\mathbf{o}$. See figure 4.16.

The origin $\mathbf{o}(t)$ is an *asymptotically stable* equilibrium if it is stable and if $|\mathbf{u}(t, t_0, \mathbf{u}^0)|$ tends to zero as $t \to \infty$ whenever $\mathbf{u}^0$ is in a particular subset of R^n containing $\mathbf{o}$. See figure 4.17.

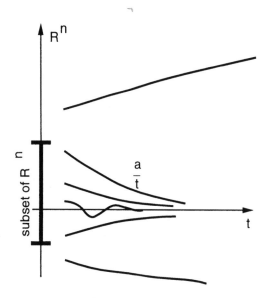

Figure 4.17: An asymptotically stable equilibrium.

The origin $\mathbf{o}(t)$ is *exponentially stable* if it is asymptotically stable and if $|\mathbf{u}(t, t_0, \mathbf{u}^0)|$ tends to zero exponentially. See figure 4.18.

Stability can also be sketched in R^n only, without the time axis. For two neurons this looks like figure 4.19.

4.6 A Lyapunov Theorem for neural networks

In this section, we answer the question whether the equilibrium $\mathbf{o}(t)$ of the fundamental system of differential equations (4.16) is stable. We will find that, under certain conditions, it is exponentially stable.

The exponential stability will be derived using Lyapunov's theorem, a most important tool in the study of stability of high dimensional dynamical systems. A.M. Lyapunov was born in 1857 in Yaroslavl, Russia. He studied at Saint Petersburg University, and published his theorem in 1892. In 1901, he became an academician at the Saint Petersburg Academy of Science. He committed suicide in 1918 after the death of his wife.

Theorem 4.4 (Lyapunov) *If there exists a positive definite decrescent C^1 function v with a negative definite derivative Dv along the solutions of the system (4.16), and if there exist functions $\phi_1, \phi_2,$ and*

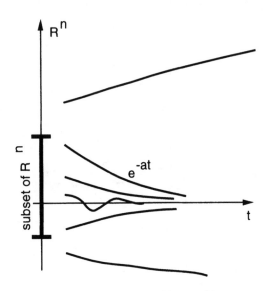

Figure 4.18: An exponentially stable equilibrium.

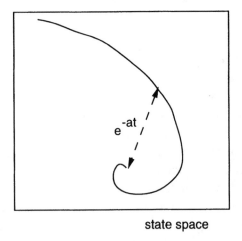

state space

Figure 4.19: Two-dimensional state space around an exponentially stable equilibrium. The distance in the figure decreases as e^{-at}, with a a constant, if t is large enough. Initially, for t close to t_0, this distance may decrease more slowly.

ϕ_3 *of the same order, such that*

$$\phi_1(|\mathbf{u}|) \leq v(\mathbf{u}, t) \leq \phi_2(|\mathbf{u}|), \tag{4.21}$$

and

$$Dv(\mathbf{u}, t) \leq -\phi_3(|\mathbf{u}|), \tag{4.22}$$

for all $|\mathbf{u}| < r$ and $t \geq t_0 \geq 0$, then the equilibrium $\mathbf{o}(t)$ of (4.16) is exponentially stable.

The function v is called a Lyapunov function. It is in general different from an energy function.

The condition $|\mathbf{u}| < r$ for some number r means that the theorem is only valid in a neighbourhood of the equilibrium $\mathbf{o}$, in a more general case, $|\mathbf{u}|$ should be replaced with the distance from the equilibrium.

A function v is called *decrescent* if there exists another function ψ, with $\psi(0) = 0$, ψ strictly increasing and continuous, such that $|v(\mathbf{u}, t)| \leq \psi(|\mathbf{u}|)$ for $|\mathbf{u}| < r$, and $t \geq t_0 \geq 0$.

The derivative Dv along a solution of (4.16) is defined by

$$Dv(\mathbf{u}) = \frac{dv(\mathbf{u})}{dt}\bigg|_{\mathbf{u} \text{ solution of (4.16)}} \tag{4.23}$$

Apart from establishing exponential stability, the Lyapunov function is also used, as an energy function, to guarantee that the network will not be stuck in a cycle, as in figure 4.20. This is related to the famous halting problem for computers, [59].

For a proof of Lyapunov's theorem, see [72]. Here, we will establish conditions on the weights and transfer functions of the network, conditions that will guarantee, via Lyapunov's theorem, the exponential stability of the equilibrium. Just as there exist no recipe for writing down an energy function, it is not possible to deduce in an automatic way the form of the Lyapunov function v or the conditions for the weights and transfer functions. Only the study of examples can help. Energy and Lyapunov functions, for example, are frequently quadratic in some transformation of the state variables.

Next, we give two conditions which will be used in establishing the exponential stability.

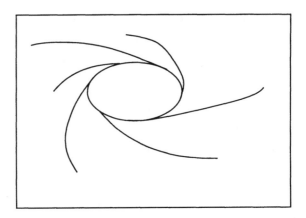

Figure 4.20: A limit cycle.

Condition 4.1 *There exist constants $r_i, i = 1, \ldots, n$, $a_{ij}, i = 1, \ldots, n$, $j = 1, \ldots, n$, such that*

$$u_i T_{ij} f_j(u_j) \le |u_i| a_{ij} |u_j|,$$

for $|u_i| < r_i, i = 1, \ldots, n$.

Here, $|u_i|$ is the absolute value of u_i.

This condition is fulfilled, for example, if $f_j(u_j) < \text{constant}|u_j|$, if $|u_j| < r_j, j = 1, \ldots, n$, see figure 4.21.

If Condition 4.1 holds, we can consider

Condition 4.2 *There exist numbers $\alpha_i > 0, i = 1, \ldots, n$, such that the matrix S with elements*

$$s_{ij} = \begin{cases} \frac{\alpha_i}{\mu_i}(-1 + a_{ii}), & i = j, \\ \frac{1}{2}\left(\frac{\alpha_i}{\mu_i}a_{ij} + \frac{\alpha_j}{\mu_j}a_{ji}\right), & i \ne j, \end{cases}$$

is negative definite, where the numbers a_{ij} are from Condition 4.1.

A matrix S is negative definite if $\mathbf{u}^T S \mathbf{u} \le 0$ for all $\mathbf{u}$, and $\mathbf{u}^T S \mathbf{u} = 0$ implies $u_i = 0, i = 1, \ldots, n$. You can check this by verifying that $-S$ is positive definite. Matrix A is positive definite if all submatrices $A_{ij}, i = 1, \ldots, k, \ j = 1, \ldots, k, \ k = 1, \ldots, n$ have positive determinant. A function $f(\mathbf{u})$ is negative definite if $f(\mathbf{u}) \le 0$ for all $\mathbf{u}$, and $f(\mathbf{u}) = 0$ implies $u_i = 0, i = 1, \ldots, n$.

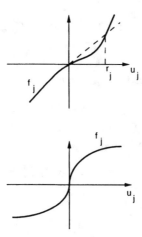

Figure 4.21: At the top of the figure is a transfer function obeying Condition 4.1. The transfer function at the bottom does not obey the condition, because its derivative is ∞ at $u_j = 0$.

Theorem 4.5 *With Conditions 4.1 and 4.2,*

$$v(\mathbf{u}) = \sum_{i=1}^{n} \frac{1}{2} \alpha_i u_i^2 \tag{4.24}$$

is a Lyapunov function for the system (4.16), and the equilibrium $\mathbf{o}$ is exponentially stable.

Proof The function $v(\mathbf{u})$ is clearly positive definite, as the numbers $\alpha_i > 0$, from Condition 4.2.

The main part of the proof consists in showing that the derivative of v along the solutions of the system is positive definite. From the definitions and Condition 4.1,

$$
\begin{aligned}
Dv(\mathbf{u}) &= \left.\frac{dv(\mathbf{u})}{dt}\right|_{\mathbf{u}\text{ solution of }(4.16)} \\
&= \left.\sum_{i=1}^{n} \alpha_i u_i \frac{du_i}{dt}\right|_{\mathbf{u}\text{ solution of }(4.16)} \\
&= \sum_{i=1}^{n} \alpha_i u_i \frac{1}{\mu_i}\left(-u_i + \sum_{j=1}^{n} T_{ij} f_j(u_j)\right)
\end{aligned}
$$

$$\leq \; \sum_{i=1}^{n} \frac{\alpha_i}{\mu_i} \left(-u_i^2 + |u_i| \sum_{j=1}^{n} a_{ij} |u_j| \right),$$
$$|u_i| < r_i, i = 1, \ldots, n. \qquad (4.25)$$

This expression is entirely in terms of the absolute values $|u_i|$, and we will abbreviate $|u_i| = w_i$. We can write this expression as $\mathbf{w}^T R \mathbf{w}$ if we choose for the elements of the matrix R

$$R_{ij} = \begin{cases} \frac{\alpha_i}{\mu_i}(-1 + a_{ii}), & i = j, \\ \frac{\alpha_i}{\mu_i} a_{ij}, & i \neq j. \end{cases} \qquad (4.26)$$

We now will simplify the condition $|u_i| < r_i, i = 1, \ldots, n$. If we define

$$r = \min(r_i), \quad i = 1, \ldots, n,$$

then $|\mathbf{u}| < r$ will imply $|u_i| < r_i, i = 1, \ldots, n$. This can easily be verified for the norms based on the Euclidean distance (4.18). From now on, we will use the Euclidean norm

$$|\mathbf{u}| = \sqrt{\sum_{i=1}^{n} u_i^2}.$$

So far, we have found the following upper bound for Dv

$$Dv(\mathbf{u}) \leq \mathbf{w}^T R \mathbf{w}, \quad |\mathbf{u}| < r. \qquad (4.27)$$

We now want to relate the matrix R to the matrix S from Condition 4.2. First consider the following equality.

$$\mathbf{w}^T R \mathbf{w} = \mathbf{w}^T \left(\frac{R + R^T}{2} \right) \mathbf{w}. \qquad (4.28)$$

This can be verified by calculating the coefficient of $w_i w_j$ in left and right hand sides. One finds

$$r_{ij} w_i w_j + r_{ji} w_j w_i = \frac{r_{ij} + r_{ji}}{2} w_i w_j + \frac{r_{ij} + r_{ji}}{2} w_j w_i.$$

If we denote by $\lambda(S)$ the largest eigenvalue of S, the matrix in Condition 4.2, and if $|\mathbf{u}| < r$,

$$Dv(\mathbf{u}) \; \leq \; \mathbf{w}^T \left(\frac{R + R^T}{2} \right) \mathbf{w}$$

$$\begin{aligned} &= \mathbf{w}^T S \mathbf{w} \\ &\leq \lambda(S)|\mathbf{w}|^2 \\ &= \lambda(S)|\mathbf{u}|^2. \end{aligned} \tag{4.29}$$

As the matrix S is negative definite, all its eigenvalues will be negative, therefore $\lambda(S)$ is negative [98]. We have now found that

$$Dv(\mathbf{u}) \leq -c_1|\mathbf{u}|^2, \quad |\mathbf{u}| < r, \tag{4.30}$$

and this is the upper bound (4.22) required for the application of the Lyapunov Theorem.

It is easy to find the bound (4.21) for $v(\mathbf{u})$. Indeed,

$$c_2|\mathbf{u}|^2 \leq v(\mathbf{u}) \leq c_3|\mathbf{u}|^2, \tag{4.31}$$

if we choose

$$c_2 = \min(\alpha_i/2), \quad c_3 = \max(\alpha_i/2), \quad i = 1, \ldots, n. \tag{4.32}$$

All the conditions for the application of Lyapunov's theorem are now fulfilled, and we can conclude, that, under Conditions 4.1 and 4.2, the equilibrium $\mathbf{o}$ of (4.16) is exponentially stable.
Q.E.D.

Remark that we did not have to assume that the weight matrix of the network was symmetric or had zero diagonal. This is in contrast with the energy function (4.10), where a symmetric weight matrix was necessary.

4.7 Problems

1. Is it a severe limitation to study only the equilibrium 0 (The function with n components that is zero everywhere) of the fundamental system of differential equations

$$\mu_i \frac{du_i(t)}{dt} = -u_i(t) + \sum_{j-1}^{n} T_{ij} f_j(u_j(t)) + I_i(t), \ i = 1, \ldots, n, \ \mu_i > 0?$$

2. How would you use a neural network in a control problem?

3. Does the fundamental system of differential equations always have to be solved numerically?

4. Can you use a single unstable equilibrium in a neural network?

5. How would you use a neural network described by a system of differential equations to retrieve data? Which parameters are important? Do you have to worry about stability of equilibria?

6. Construct a neural network for which Conditions 4.1 and 4.2 hold. Find the equilibria by simulation.

4.8   Project: Chaos

The two attractors sketched on the right hand side of figure 4.13 are *chaotic*, because the behaviour of a trajectory attracted to such an attractor is totally unpredictable in the long term. The aim of this project is to make a numerical study of chaos in neural networks.

The study of chaos is easy and difficult at the same time. Once you adopt a dynamical system as (4.2) as a tool to model time-dependent behaviour, it is easy to observe trajectories that look chaotic. Examples are [44] the evolution of the weather, the orbits of the planets, and electro-encephalograms. To prove the critical dependency of the trajectories on initial conditions, is much more difficult, and requires extensive mathematics [95].

If you observe a system which looks chaotic, it is difficult to determine whether the system is noisy, or whether the system is deterministic, and chaotic. When observing firing in neurons, for example, it is known that part of the observed changes in membrane potential are caused by the deterministic dynamics (4.1). But there is also measurement noise, thermal noise, electrochemical noise, etc. The measurements look chaotic, and it is not known whether this apparent chaos is entirely due to noise, or whether the dynamics (4.1) are chaotic [84, 41].

Consider a two-neuron system, with u and v the states of the two neurons,

$$\dot{u} = -u + \tanh u + \tanh v, \qquad (4.33)$$
$$\dot{v} = -v - 100 \tanh u + 0.75 \tanh v. \qquad (4.34)$$

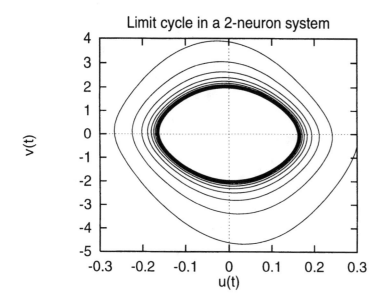

Figure 4.22: A periodic attractor in the two-neuron system (4.34).

Discretize this system using the Runge-Kutta method (4.5), or the simple Euler method [62], available in a number of numerical software packages. Plot some trajectories for an appropriately small stepsize. You will find a periodic attractor as in figure 4.26

Now add a term $0.3\cos(t)$ to the right hand side of equation (4.34). Plot trajectories with the same initial states as before. You will see something like in figure 4.27. If you plot the trajectory over a long enough time interval, you will see a black region in the graph. This may indicate that the trajectory does never come back in the same point of state space, and completely fills a region. To prove this would be very difficult.

Instead of the cosine, add some noise, uniformly distributed between -1 and 1. Again plot some trajectories. If you compare this plot with the previous one, would you be able to tell which one has the cosine, and which one has the noise?

Most chaotic attractors studied are in two, three, or four dimensions. Systems with only a few neurons are very limited, however. Add extra equations to (4.34), so that you obtain a ten-neuron system. Integrate this with the Euler method. Plot the evolution in time of the first two neurons. This is a two-dimensional projection of the 10-

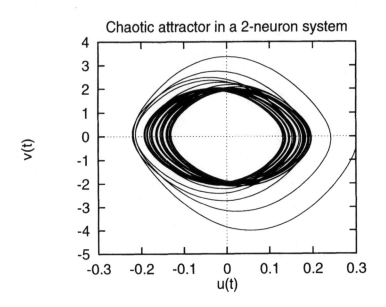

Figure 4.23: A chaotic attractor in the two-neuron system (4.34).

dimensional state space. Do you still find a periodic attractor? Add a periodic input, such as cosine, to some of the neurons. Do you observe chaotic attractors? Try also adding noise instead of a periodic input.

The neurons in the system (4.34) have some direct feedback, i.e. u and v appear in the right hand side in terms other than just $-u$ and $-v$. In other words, the synapse matrix T_{ij} has non-zero diagonal. Choose a zero-diagonal weight matrix, and see whether you can find a system that still has a chaotic attractor. This difficult problem is made easier if you choose an anti-symmetric weight matrix. We return to the problem of the diagonal of the weight matrix on pages 122 and 142.

Chaos can be a good model of some rare phenomena, for example epilepsy, or irregular heartbeat, but in most engineering applications, where you want to use attractors to store patterns, it is an unwanted feature. Some researchers [86] try to *control* chaos. This usually consists in forcing a system out of a chaotic attractor by using external inputs. The inputs I_i in (4.1) are a convenient way to do this. Try it. Use your previous network with a periodic attractor, add a periodic perturbation that gives you a chaotic attractor. Add inputs to get the system out of the attractor. Does it go to another attractor? How does the magnitude of the inputs compare to that of the periodic perturbation?

There are many more facets to chaos. The publications [92, 3] will be at your level.

Chapter 5

Synchronous and Discrete Networks

In chapter 3 and 4, we have studied networks in continuous time. This is not easy to implement with electronic circuits, certainly not for large networks. In addition to this, most electronics is digital, and it is difficult to implement sufficiently smooth transfer functions in digital electronics.

In this chapter, we will construct networks with binary transfer functions, and in discrete time. We will obtain this by successive simplifications of the networks studied so far. We will have to derive the energy function again, and we will also investigate what can happen in the state space.

Chapters 6 and 7 will exclusively treat the number of attractors in the state space of the neural network, and chapter 8 will be about the effective use of noise in the network.

This chapter is in many ways a link between the first and second part of the book. It connects the continuous transition functions from chapters 2, 3, and 4 with the discrete ones in the following chapters. It leads from the analog electronics in chapter 4 to digital electronics, but with -1 and 1 states instead of 0 and 1. It abandons the idea of continuous time for two new modes of operation, asynchronous and synchronous.

5.1   From Analog to Digital Neural Networks

Let us reconsider the fundamental system of differential equations, studied in chapter 3 and 4

$$\mu_i \frac{du_i(t)}{dt} = -u_i(t) + \sum_{j=1}^{n} T_{ij} f_j(u_j(t)) + I_i(t), \quad i = 1, \ldots, n, \quad \mu_i > 0.$$
(5.1)

The digital form of neural networks is usually derived from another set of equations, very similar to (5.1), apart from the fact that the function f_j is now in front of the summation sign [75],

$$\frac{dx_i(t)}{dt} = -x_i(t) + f_i \left(\sum_{j=1}^{n} T_{ij} x_j(t) - \theta_i \right), \quad i = 1, \ldots, n.$$
(5.2)

The thresholds $\theta_i, i = 1, \ldots, n$, are independent of time.

In order to relate the set of equations (5.2) to (5.1), consider the transformation

$$w_i = \sum_{j=1}^{n} T_{ij} x_j, \quad i = 1, \ldots, n,$$
(5.3)

and assume that the determinant of the synapse matrix is nonzero, so that the transformation is invertible,

$$\det T \neq 0.$$
(5.4)

If the left and right hand sides of (5.2) are multiplied with T_{ij} and summed over j, one obtains

$$\sum_{j=1}^{n} T_{ij} \frac{dx_j(t)}{dt} = -\sum_{j=1}^{n} T_{ij} x_j(t) + \sum_{j=1}^{n} T_{ij} f_j \left(\sum_{k=1}^{n} T_{jk} x_k(t) - \theta_j \right),$$
$$i = 1, \ldots, n.$$
(5.5)

or,

$$\frac{dw_i(t)}{dt} = -w_i(t) + \sum_{j=1}^{n} T_{ij} f_j(w_j - \theta_j)$$
(5.6)

We can now perform the transformations $u_i = w_i - \theta_i$, $\theta_i = -I_i$, and $t' = \mu t, \mu > 0$. The last transformation is a change in clock speed.

Equation (5.6) now becomes

$$\mu\frac{du_i(t)}{dt} = -u_i(t) + \sum_{j=1}^{n} T_{ij}f_j(u_j(t)) + I_i(t), \quad i = 1,\ldots,n, \quad \mu > 0,$$
(5.7)

which is entirely similar to (5.1), apart from the fact that the time constant or decay rate μ is now the same for all neurons.

The conclusion up to now is that the dynamics (5.2) is equivalent to the dynamics (5.1) if the time constants μ_i are the same for all neurons, the determinant of the synapse matrix is nonzero, and the external inputs I_i are independent of time.

It is now easy to discretize the time in (5.2). Use the approximation

$$\frac{dx_i(t)}{dt} = x_i(t+1) - x_i(t),$$
(5.8)

and (5.2) can be written as

$$x_i(t+1) = f_i\left(\sum_{j=1}^{n} T_{ij}x_j(t) - \theta_i\right), \quad i = 1,\ldots,n.$$
(5.9)

The digital version of the network is now obtained by letting the slope of f_i in the origin become infinite. The dynamics of the network are then described by

$$x_i(t+1) = \text{sgn}\left(\sum_{j=1}^{n} T_{ij}x_j(t) - \theta_i\right), \quad i = 1,\ldots,n,$$
(5.10)

with

$$\text{sgn}\,x = \begin{cases} 1, & x \geq 0, \\ -1, & x < 0. \end{cases}$$
(5.11)

The sgn function is a threshold function, and is studied in detail in Chapter 6. The sgn function can also be mapped into a Boolean function, this is studied in section 7.2.

The network dynamics (5.10) are particularly easy to understand and to implement on chip. This is one of the reasons for the success of neural networks. Despite the simplifications, they have retained much of the rich dynamical behaviour of the nonlinear systems (5.1). This is the subject of the rest of this book.

5.2   Synchronous Dynamics and Cycles

A network operating in discrete time can function in two modes. They are called *synchronous* and *asynchronous*.

In asynchronous mode, the equation

$$x_i(t+1) = \text{sgn}\left(\sum_{j=1}^{n} T_{ij}x_j(t) - \theta_i\right) \tag{5.12}$$

is applied for a neuron i chosen randomly for each time step. This random updating makes the analysis quite difficult. There is no simple answer, for example, to the problem when to stop updating the neurons. This opens up a whole unexplored field of research.

In the sequel, we will assume synchronous updating, i.e. the set of equations

$$x_i(t+1) = \text{sgn}\left(\sum_{j=1}^{n} T_{ij}x_j(t) - \theta_i\right), \quad i = 1, \ldots, n, \tag{5.13}$$

is applied for all i, so that all neurons are updated every time step.

To investigate the convergence behaviour of the network (5.13), we will need to construct an energy function, as in Chapter 4. There is no recipe for an energy function, and we make the following choice

$$
\begin{aligned}
E(\mathbf{x}(t)) &= -\sum_{i=1}^{n} x_i(t) \sum_{j=1}^{n} T_{ij}x_j(t-1) + \sum_{i=1}^{n} \theta_i[x_i(t) + x_i(t-1)], \\
&= -\mathbf{x}^T(t)T\mathbf{x}(t-1) + \theta^T[\mathbf{x}(t) - \mathbf{x}(t-1)], \tag{5.14}
\end{aligned}
$$

where $\mathbf{x}^T$ is the transpose of the column vector $\mathbf{x}$.

In the definition (5.11) of the function sgn, the value 0 has been arbitrarily assigned a sgn equal to 1. This arbitrariness is a very deep problem in the theory of neural networks. For networks in discrete time, we will suffice with

Theorem 5.1 *In the network (5.13), the thresholds can always be adjusted so that the argument of the sgn function is never zero, and without disturbing the trajectories of the network.*

Proof. Assume that for certain values of the state vector $\mathbf{x}$ and for certain neurons i the argument $\sum_j T_{ij}x_j - \theta_i$ of the sgn function is zero. As there are 2^n states and n neurons, it is possible to calculate the smallest value of the absolute value of the argument that is non-zero. This value can be mathematically expressed as

$$\min_{\mathbf{x},i} \left| \sum_j T_{ij}x_j - \theta_i \right|, \quad \sum_j T_{ij}x_j - \theta_i \neq 0.$$

Call this value ϵ.

If we now subtract $\epsilon/2$ from all thresholds $\theta_i, i = 1, \ldots, n$,

$$\theta_i' = \theta_i - \epsilon, \quad i = 1, \ldots, n,$$

then

$$\mathrm{sgn}\left(\sum_j T_{ij}x_j - \theta_i' \right) = \mathrm{sgn}\left(\sum_j T_{ij}x_j - \theta_i \right),$$

as all arguments $\sum_j T_{ij}x_j - \theta_i$ of the sgn function will keep the same sign. The arguments $\sum_j T_{ij}x_j - \theta_i$ that were zero and had a sgn 1, are now equal to $\epsilon/2$, and still have a sgn equal to 1. This proves that nothing has changed to the dynamics of the network.
Q.E.D.

We are now ready to prove that the function (5.14) is indeed an energy function, i.e.

Theorem 5.2 *The energy*

$$E(\mathbf{x}(t)) = -\sum_i x_i(t) \sum_j T_{ij}x_j(t-1) + \sum_i \theta_i[x_i(t) + x_i(t-1)]$$

is decreasing during the synchronous operation of the network (5.13), if T is symmetric.

Proof. For a symmetric synapse matrix T,

$$
\begin{aligned}
E(\mathbf{x}(t)) \quad &- \quad E(\mathbf{x}(t-1)) \\
&= \quad -\sum_i x_i(t) \sum_j T_{ij}x_j(t-1) + \sum_i \theta_i[x_i(t) + x_i(t-1)] \\
&+ \quad \sum_i x_i(t-1) \sum_j T_{ij}x_j(t-2) - \sum_i \theta_i[x_i(t-1) + x_i(t-2)]
\end{aligned}
$$

$$= -\sum_i x_i(t) \sum_j T_{ij}x_j(t-1) + \sum_i [x_i(t) - x_i(t-2)]\theta_i$$

$$+ \sum_j x_j(t-1) \sum_i T_{ij}x_i(t-2)$$

$$= -\sum_i [x_i(t) - x_i(t-2)] \left[\sum_j T_{ij}x_j(t-1) - \theta_i \right]. \qquad (5.15)$$

We have to establish the sign of this expression. Assume i is fixed, equivalent to choosing one term from the sum over i in (5.15). Listing all possibilities for this term, we can establish the following table

$x_i(t-2)$	$x_i(t-1)$	$x_i(t)$	$x_i(t-2) - x_i(t)$	$\mathrm{sgn}\left(\sum T_{ij}x_j(t-1) - \theta_i\right)$
-1	-1	-1	0	-1
-1	-1	1	-2	1
-1	1	-1	0	-1
-1	1	1	-2	1
1	-1	-1	2	-1
1	-1	1	0	1
1	1	-1	2	-1
1	1	1	0	1.

In this table, $\mathrm{sgn}\left(\sum_j T_{ij}x_j(t-1) - \theta_i\right)$ was derived from the transition $x_i(t-1) \to x_i(t)$. It was also assumed that the thresholds were adapted so that the argument of sgn was never zero. That this is possible was proven in Theorem 5.1.

It follows from the table that

$$-[x_i(t) - x_i(t-2)] \left(\sum_j T_{ij}x_j(t-1) - \theta_i \right) \le 0, \quad i = 1, \dots, n,$$

so that we can conclude

$$E(\mathbf{x}(t)) - E(\mathbf{x}(t-1)) \le 0,$$

i.e., E is an energy function for the network (5.13).
Q.E.D.

Using the previous theorem, it is possible to derive a result about the trajectories of the network (5.13). It turns out that, for synchronous operation, the trajectory either converges to a single state, or oscillates between two states. The fact that long cycles are not possible is important in engineering design. The user of the network wants to store attractors consisting of a single state, and does not want the network to end up in a limit cycle, hesitating between several states. If the network oscillates between two states, these can be shown to the user as alternatives. This is not possible for a long limit cycle.

Theorem 5.3 *The period of a synchronously operating neural network (5.13) is 1 or 2, if T is symmetric.*

Proof. Assume that the states $\mathbf{x}(0), \mathbf{x}(1), \ldots, \mathbf{x}(p-1)$ form a cycle of period p. Then, the energy E is constant for all states in the cycle,

$$E(\mathbf{x}(0)) = E(\mathbf{x}(1)) = \ldots = E(\mathbf{x}(p-1)). \tag{5.16}$$

If $p > 2$, then $\mathbf{x}(2) \neq \mathbf{x}(0)$. In the table in the proof of Theorem 5.2, single out the rows for which $x_i(t) \neq x_i(t-2)$. For the remaining rows, you can see that the energy is strictly decreasing, or

$$E(\mathbf{x}(1)) - E(\mathbf{x}(0)) < 0.$$

This contradicts (5.16), so that we conclude that the period $p \leq 2$. Q.E.D.

There exist similar results for non-symmetric matrices. They are analyzed with the same techniques used for symmetric synapse matrices [45].

5.3 Problems

1. If the matrix T is symmetric and has zero diagonal, is its determinant always different from zero?

2. Is it possible to circumvent the restriction that the time constants have to be equal when establishing the equivalence of the neural networks (5.1) and (5.2)? (This is a difficult problem)

3. Is the discretization error in (5.8) somehow dependent on the time constant μ?

4. What are the essential differences between an analog and a discrete neural network model? Discuss the advantages and disadvantages of both.

5. Are all equilibria stable in a network with discrete neuron states and discrete time?

6. Can a limit cycle occur in a discrete time, discrete state network?

7. Consider a neural network with 2 neurons and a weight matrix with non-diagonal elements equal to -1. Describe the trajectories in state space for synchronous and asynchronous updating. Choose an energy function and verify that the energy does not increase during any state transition. Construct an equivalent Boolean network.

5.4   Project: Continuum of Neurons

We have encountered up to now neurons with continuous or discrete states, evolving in continuous or discrete time. The neurons were always distinct entities, numbered by $i = 1, \ldots, n$. Some experiments show that in the brain, waves of activity can travel through the neurons. A number of neurons that are close to each other, are active at the same time. The wavelength of these waves is a few millimeters [84]. This is large compared to the size of the neuron soma, a few tens of microns in size. Drawn on the scale of the wave, the neurons are so small that they appear as a continuum of neurons.

To model this, we need to indicate where a neuron is located. We will use a three-dimensional vector $\mathbf{y}$ for this. The origin and orientation of the coordinate axes will depend on the feature of the brain that is studied, in the cortex for example, one axis is often perpendicular to the skull, and the two others tangent to the skull. The state of the neurons will be indicated by $I(\mathbf{y}, t)$, a function of the location of the neuron, and the time. The neurons are distributed in space with a density $\rho(\mathbf{y})$. This density is the number of neurons per unit of volume, in the limit for small volume elements.

A function $f(I(\mathbf{y}, t))$ will indicate whether the neuron at location $\mathbf{y}$ is active or not. The function f is similar to the transfer function.

The weights are now a function $w(\mathbf{y}, \mathbf{y}')$, indicating the strength of the connection from the neuron at location $\mathbf{y}'$ to the neuron at $\mathbf{y}$.

The input to the neuron at location $\mathbf{y}$ will be $w(\mathbf{y}, \mathbf{y}')f(I(\mathbf{y}', t))$, summed over all neurons at $\mathbf{y}'$. Because space is continuous, this summation is an integral. In this summation, we take account of the density of the neurons $\rho(\mathbf{y'})$. In discrete time, the dynamics of the neural network are

$$I(\mathbf{y}, t+1) = \int_{-\infty}^{\infty} w(\mathbf{y}, \mathbf{y'}) f(I(\mathbf{y'}, t)) \rho(\mathbf{y'}) \, d\mathbf{y'}. \tag{5.17}$$

Stability will be reached when

$$I(\mathbf{y}, t) = \int_{-\infty}^{\infty} w(\mathbf{y}, \mathbf{y'}) f(I(\mathbf{y'}, t)) \rho(\mathbf{y'}) \, d\mathbf{y'}. \tag{5.18}$$

This is a complex integral equation for $I(\mathbf{y}, t)$.

Often the weights are only a function of the vector pointing from $\mathbf{y'}$ to $\mathbf{y}$, and the equation for I becomes

$$I(\mathbf{y}, t+1) = \int_{-\infty}^{\infty} w(\mathbf{y} - \mathbf{y'}) f(I(\mathbf{y'}, t)) \rho(\mathbf{y'}) \, d\mathbf{y'}. \tag{5.19}$$

This is nothing else than the convolution [47] of w and $f(I)\rho$.

To simplify the matter, assume that all neurons are on a line, with their position indicated by the scalar y. The dynamics are then

$$I(y, t+1) = \int_{-\infty}^{\infty} w(y, y') f(I(y', t)) \rho(y') \, dy'. \tag{5.20}$$

Try using Leibnitz's rule for differentiation under the integral sign (3.4). Does it make the equation any simpler?

In order to get a grasp on this equation, we will have to do drastic simplifications. Assume that the density ρ is 1 everywhere. Also assume that the weights are only non-zero for neighbouring neurons at most a distance a apart,

$$w(y, y') = w(|y - y'|) = \begin{cases} 1, & |y - y'| \leq a, \\ 0, & |y - y'| > a. \end{cases} \tag{5.21}$$

The integral in (5.20) is now a convolution, and can be simplified to (do this yourself)

$$I(y, t+1) = \int_{y-a}^{y+a} f(I(y', t)) \, dy'. \tag{5.22}$$

In figure (5.1) is an illustration of this, for

$$I(y', 0) = \frac{1}{3}(\sin(3y') + \cos(y'/2) + \tanh(y'))$$

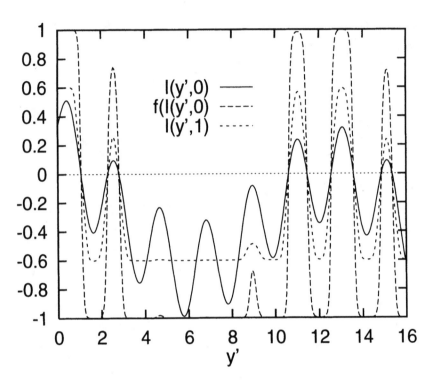

Figure 5.1: $I(y',0) = \frac{1}{3}(\sin(3y') + \cos(y'/2) + \tanh(y'))$, the state function of the neurons, its nonlinear transformation $f(I(y',0)) = \tanh(10I(y',0))$, and the state function one time step later, $I(y',1)$.

and

$$f(I(y',t)) = \tanh(10I(y',t)).$$

The choice of a steep tanh for f can be biologically motivated as a transfer function, but the choice of $I(y',0)$ is random.

Program the equation (5.22) yourself. Choose an initial state function $I(y',0)$, and a window size a for the convolution. Calculate $I(y',t)$ for large values t of the time. Is an equilibrium reached? What does the convolution do to the frequencies that compose the initial state function $I(y',0)$?

Now, perturb the function $I(y',0)$ by noise with nearly the same amplitude. The function will become discontinuous, which presents a problem both in theory and in practice. In theory, a function which jumps from one value to another for values of its argument y' that can be infinitely close, need a special definition for its integral, the Lebesgue integral [11]. We will not investigate this topic here.

The practical problem resides in the numerical integration of the left hand side of (5.22). A typical numerical integration algorithm is Simpson's rule [4],

$$\int_{x_0}^{x_{2n}} f(x)\,dx = \frac{h}{3}[f_0 + 4(f_1 + f_3 + \ldots + f_{2n-1})$$

$$+ 2(f_2 + f_4 + \ldots + f_{2n-2}) + f_{2n}] - \frac{nh^5}{90}f^{(4)}(\xi),$$

$$f_i = f(x_i) = f(x_0 + ih), \quad x_0 \le \xi \le x_{2n}, \qquad (5.23)$$

where $f^{(4)}$ is the fourth derivative of f. If the function f fluctuates strongly, it wil have a large fourth derivative, and it will be necessary to choose h small enough so that the error term $nh^5/90\,f^{(4)}(\xi)$ in (5.23) is small enough.

A continuum of neurons is useful whenever waves of neural activity have to be modelled. If the states of the individual neurons cannot be measured, it is often possible to measure the electric or magnetic fields generated by the waves of activity. This is used in electro encephalography (EEG), magneto encephalography (MEG), and functional magnetic resonance imaging (fMRI). It is important to realize that these techniques measure properties of groups of neurons, hence the measured quantities are averages. The activities of the individual neurons can be very different from these averages.

Explore this in your simulations. Generate an initial state function with high frequencies. Are these frequencies still present in the equilibrium? Try the convolution weight function (5.21), but also other weight functions that do depend on the position of the neurons, not just their distance.

Chapter 6

Linear Capacity

In this chapter and the following, we will concentrate exclusively on the number of attractors or equilibria that can be stored in the network. This number is called the *capacity* of the network, and it is one of the most important characteristics of a network. If a neural network is used as an associative memory, the first property the user will want to know is how much can be stored in the memory.

There are several ways to define capacity, dependent on what sort of stability is required for the equilibria. In this book, we follow an approach which we think is most useful for engineers [5]. The reader should know that there exists a whole parallel field of investigation into the capacity and many more aspects of neural networks, based on the statistical physics of spin glasses [10].

In this chapter, we will prove one result, namely that the capacity is linearly proportional to the number of neurons. In preparing for this result, we will encounter several concepts and techniques which are useful in their own right. Hyperplanes, for example, are also used in the analysis of multilayer feed-forward networks. We will also compare the number of threshold functions with the number of Boolean functions. This is a classical result in engineering mathematics, and is useful in the comparison of networks with threshold nodes and networks with Boolean nodes [6]. The linear capacity result in itself is a very general result, encompassing the more specific results that will be derived in chapter 7, and some results from spin glasses.

6.1   Threshold Functions

In order to calculate the capacity of a network, we need to know how many really different networks there are. If you multiply all weights in the network

$$x_i(t+1) = \text{sgn}\left(\sum_{j=1}^{n} T_{ij}x_j(t) - \theta_i\right), \quad i = 1, \ldots, n, \qquad (6.1)$$

by the same positive number, for example, the new network will have the same attractors and the same trajectories as the old one. In order to clarify the equivalence of networks, we will study the functions

$$\text{sgn}\left(\sum_{j=1}^{n} T_{ij}x_j - \theta_i\right) \qquad (6.2)$$

in some detail.

A function f with values 1 and -1 is called a *threshold function* if the inverse image of 1, $f^{-1}(1)$, and of -1, $f^{-1}(-1)$, are separated by a hyperplane.

The geometric space we work in is the state space, and because only the sgn function is used, this space is reduced to the corners of the hypercube [15]. For n neurons, we denote this state space by $\{-1, +1\}^n$. It consists of vectors with n coordinates, all equal to +1 or -1. The space $\{-1, +1\}^n$ is sometimes called n-dimensional binary space, or n-dimensional bipolar space.

A hyperplane π in this space has equation

$$\mathbf{a} \cdot \mathbf{x} = \theta, \quad \mathbf{x} \in \{-1, +1\}^n, \quad \mathbf{a} \in R^n.$$

This concept is illustrated for one dimension in figure 6.1, and for two dimensions in figure 6.2.

The threshold function f does not have to defined on all 2^n points of $\{-1, +1\}^n$. We will understand this better if we can calculate B_n^m, *the number of threshold functions of n binary variables defined on m points.*

In order to calculate B_m^n, we will augment the dimension of the space, so that all hyperplanes go through the origin. Instead of the n variables $x_1, \ldots, x_n$, we consider the $n+1$ variables $x_1, \ldots, x_n, \theta$, and

$$\mathbf{a} \cdot \mathbf{x} - \theta = 0, \quad (\mathbf{x}, \theta) \in \{-1, +1\}^n \times R \qquad (6.3)$$

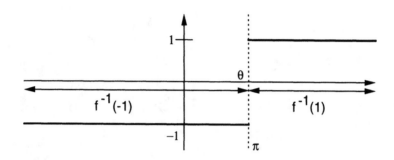

Figure 6.1: A threshold function in one dimension.

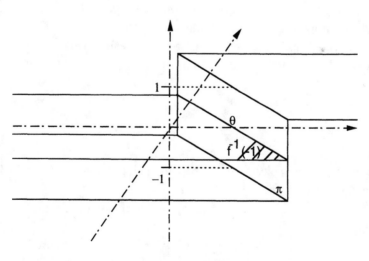

Figure 6.2: A threshold function in two dimensions.

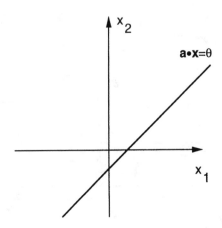

Figure 6.3: A hyperplane in two dimensions is a straight line.

is the equation of a hyperplane in $n+1$ dimensions, going through the origin, because $x_i = 0$, $i = 1,\ldots,n$, $\theta = 0$ is a solution of (6.3). See figure 6.3.

Assume now that the threshold function f is defined on m points $\mathbf{x}^1,\ldots,\mathbf{x}^m$. These are points in n-dimensional space. Up to now, in the equation $\mathbf{a}\cdot\mathbf{x} - \theta = 0$, we have considered $\mathbf{x}$ and θ to be the variables. We will now consider $a_1,\ldots,a_n,\theta$ to be the variables. This is motivated by the fact that, further on in the calculations, m will be the number of equilibria, and $\mathbf{x}^1,\ldots,\mathbf{x}^m$ equilibria or patterns will be given by the user, and fixed.

The set of equations

$$\mathbf{a}\cdot\mathbf{x}^i - \theta = 0, \quad i = 1,\ldots,m \tag{6.4}$$

defines m hyperplanes in $n+1$-dimensional space. Refer to figure 6.4 for an illustration.

If you take into account that the lines drawn in figure 6.4 are hyperplanes separating the inverse images of $+1$ and -1 under the threshold function, it becomes clear that two regions as indicated in figure 6.4 correspond to different functions because, if $\mathbf{x}^i$ as a point is mapped to $+1$ in one region, it is mapped to -1 in the other region.

Remember from elementary geometry that a line $\mathbf{a}\cdot\mathbf{x} - \theta = 0$ separates a plane in two halves, with $\mathbf{a}\cdot\mathbf{x} - \theta > 0$ for all points in one half plane, and $\mathbf{a}\cdot\mathbf{x} - \theta < 0$ for the points in the other half plane.

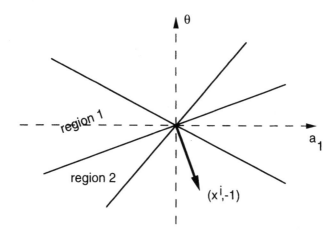

Figure 6.4: Region 1 and region 2 correspond to different threshold functions.

Again referring to figure 6.4, all values (a_1, θ) in the same region correspond to the same function. Two points in different regions correspond to different functions as at least one $\mathbf{x}^\alpha$ of the m vectors $\mathbf{x}^1, \ldots, \mathbf{x}^m$ is mapped to +1 by one function and to -1 by the other.

What we have established here is a link between the number of regions in a space and the number of threshold functions. More precisely, if C_{n+1}^m is the maximum number of regions in $n + 1$-dimensional space, made by m hyperplanes through the origin, then

$$B_n^m \le C_{n+1}^m. \tag{6.5}$$

We will now try to calculate C_{n+1}^m. First, we will establish a recursion equation, and then solve this equation. For clarity, this is formulated a two theorems.

Theorem 6.1 *If C_{n+1}^m is the maximum number of regions in $n + 1$-dimensional space, made by m hyperplanes through the origin, then*

$$C_{n+1}^m = C_n^{m-1} + C_{n+1}^{m-1}.$$

Proof. Consider C_{n+1}^{m-1}, the number of regions made by $m - 1$ hyperplanes. Now add an m-th hyperplane to make the most possible number of new regions. Refer to figure 6.5.

The m-th plane intersects the $m - 1$ planes in at most $m - 1$ hyperlines. Hyperlines have one dimension less than hyperplanes. In three

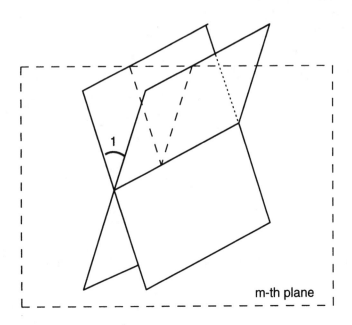

Figure 6.5: How the m-th plane intersects $m - 1$ planes in at most $m - 1$ hyperlines.

dimensions, for example, the hyperplanes are planes, and the hyperlines are lines.

The hyperlines in $n + 1$-dimensional space are hyperplanes in n-dimensional space. The m-th plane itself is a space of n dimensions, containing the $m - 1$ hyperlines. We can now use the definition of the numbers C_{n+1}^{m-1}, to conclude that the $m - 1$ hyperlines divide the m-th plane in at most C_n^{m-1} regions.

In figure 6.5, you can see that region 1 has been split in two by the m-th plane. In general, in the $n + 1$ dimensional space, one region is added per region in the m-th plane. As there are at most C_n^{m-1} such regions, we find the following recursion equation

$$C_{n+1}^m = C_n^{m-1} + C_{n+1}^{m-1}. \tag{6.6}$$

Q.E.D.

We now have to solve this recursion equation.

Theorem 6.2 *If C_{n+1}^m is the maximum number of regions in $n + 1$-*

dimensional space, made by m hyperplanes through the origin, then

$$C_{n+1}^m = 2 \sum_{i=0}^{n} \binom{m-1}{i}.$$

Proof. The recursion equation (6.6) is a two-dimensional difference equation, the two independent variables being n and m. Just as for a two-dimensional partial differential equation, we will need to find boundary conditions in both variables.

For $n = 0$, C_{n+1}^m is about m hyperplanes in dimension one. These hyperplanes are points. These points divide the one-dimensional space, a line, in two half lines, so that

$$C_1^m = 2.$$

For $n = 1$, the hyperplanes are lines in a two dimensional space, so that

$$C_2^m = 2m, \tag{6.7}$$

what can be easily verified by drawing m lines through the origin and observing that they divide the plane in $2m$ regions. These are boundary conditions in the variable n.

As the maximum number of regions in a one-dimensional space is always 2,

$$C_{n+1}^1 = 2,$$

providing a boundary condition in m.

Having established boundary conditions, we need to find a general solution to the recursion equation (6.6). This is usually found with the help of generating functions [49], but we will use a more intuitive approach.

First, observe that the binomial coefficient $\binom{m}{n+1}$ is a solution of (6.6). But then, $\binom{m-1}{n}$ is also a solution. We will therefore look for a general solution of the form

$$C_{n+1}^m = \sum_{p,q} a_{pq} \binom{m-p}{n+1-q}. \tag{6.8}$$

Using the initial conditions and the recursion equation (6.6), we can make a table of C_{n+1}^m. For simplicity, we will list $C_{n+1}^m/2$, and we will also list the differences between $C_{n+1}^m/2$ and $C_n^m/2$.

$m \downarrow n \rightarrow$	0	1		2		3		4
1	1	1		1		1		1
2	1	1	2	2		2		2
3	1	2	3	1	4	4		4
4	1	3	4	3	7	1	8	8
5	1	4	5	6	11	4	15	1	16
6	1	5	6	10	16	10	26	5	31

Observe now that the differences form the table of binomial coefficients! This means

$$C_{n+1}^m = C_n^m + 2 \binom{m-1}{n}. \tag{6.9}$$

Combined with the boundary condition $C_{n+1}^1 = 2$, this gives

$$C_{n+1}^m = 2 \sum_{i=0}^{n} \binom{m-1}{i}. \tag{6.10}$$

Q.E.D. For more background to these calculations, refer to [87].

We can now combine the inequality (6.5) and (6.10) into

Theorem 6.3 *If B_n^m is the number of threshold functions of n binary variables defined on m points, then*

$$B_n^m \le 2 \sum_{i=0}^{n} \binom{m-1}{i}. \tag{6.11}$$

6.2   Linear Capacity

The equations governing the operation of the network we are studying are

$$x_i(t+1) = \operatorname{sgn}\left(\sum_{j=1}^{n} T_{ij} x_j(t) - \theta_i \right), \quad i = 1, \ldots, n, \tag{6.12}$$

with

$$\operatorname{sgn} x = \begin{cases} 1, & x \ge 0, \\ -1, & x < 0. \end{cases}$$

Because the sum $\sum_{j=1}^{n} T_{ij}x_j(t) - \theta_i)$ is squashed by the sgn function, the state vector lies on one of the corners of the hypercube $\{-1, +1\}^n$.

The state $\mathbf{x}$ is a *fixed vector*, or *equilibrium*, or *stored pattern*, if

$$x_i(t) = \operatorname{sgn}\left(\sum_{j=1}^{n} T_{ij}x_j(t) - \theta_i\right), \quad i = 1, \ldots, n. \tag{6.13}$$

We want to find how many such fixed vectors $\mathbf{x}$ there can be, given a number n of neurons, and given the freedom to adjust the synapses T and the thresholds $\theta_i, i = 1, \ldots, n$.

Theorem 6.4 *If for every set of $m < 2^{n-1}$ vectors in $\{-1, +1\}^n$ one has to be able to find a zero-diagonal $n \times n$ matrix T and thresholds $\theta_i, i = 1, \ldots, n$, such that the m vectors are fixed vectors, then $m \leq n$.*

Proof. In the state space $\{-1, +1\}^n$, construct m vectors

$$\mathbf{x}^\alpha = (x_1^\alpha, x_2^\alpha, \ldots, x_m^\alpha, x_{m+1}^\alpha, \ldots, x_n^\alpha). \tag{6.14}$$

For every α, $x_2^\alpha, \ldots, x_m^\alpha, x_{m+1}^\alpha, \ldots, x_n^\alpha$ are chosen fixed, such that all $\mathbf{x}^\alpha$ are different. This is possible because we assumed $m < 2^{n-1}$.

If all the $\mathbf{x}^\alpha, \alpha = 1, \ldots, m$ have to be fixed vectors, then, for any choice of the first coordinates $x_1^1, x_1^2, \ldots, x_1^m$, there must exist numbers $T_{12}, T_{13}, \ldots, T_{1n}$, and θ_1 such that

$$x_1^\alpha = \operatorname{sgn}\left(\sum_{j=2}^{n} T_{1j}x_j^\alpha - \theta_1\right), \quad \alpha = 1, \ldots, m. \tag{6.15}$$

Remember that $T_{11} = 0$, so that the sum starts from $j = 2$.

The m first coordinates $x_1^1, x_1^2, \ldots, x_1^m$ of the fixed vectors can be chosen in 2^m ways. For each choice, we must find a threshold function (6.15) of $n - 1$ variables

$$x_2^\alpha, x_3^\alpha, \ldots, x_n^\alpha$$

defined in m points $\alpha = 1, \ldots, m$.

This means that $B_{n-1}^m \geq 2^m$ must hold. But from (6.11),

$$B_{n-1}^m \leq 2 \sum_{i=0}^{n-1} \binom{m-1}{i},$$

so that we must have

$$2 \sum_{i=0}^{n-1} \binom{m-1}{i} \geq 2^m. \tag{6.16}$$

We always assume that the binomial coefficients are zero when the parameters are out of range, e.g.

$$\binom{a}{b} = 0, \quad a < b.$$

Suppose $m > n$, then (6.16) implies

$$\sum_{i=0}^{n-1} \binom{m-1}{i} < \sum_{i=0}^{m-1} \binom{m-1}{i} = 2^{m-1}. \tag{6.17}$$

From (6.16) and (6.17) it would now follow that $2^m > 2^m$, a contradiction, so that our assumption must have been false, and $m \leq n$. Q.E.D.

The requirement that T be zero-diagonal is a very severe one. Indeed, if T was the identity matrix, more or less the opposite of a zero-diagonal matrix, then condition (6.13) becomes

$$x_i(t) = \text{sgn}(x_i(t) - \theta_i), \quad i = 1, \ldots, n,$$

which is trivially fulfilled for $\theta_i = 0, i = 1, \ldots, n$. This means that, if T is the identity matrix, the capacity $m = 2^n$, exponential in the number of neurons. This is true for many variants of neural networks. If the capacity is superlinear, the diagonal of T is usually non-zero.

To calculate the capacity, one could also look for the eigenvalues of T. But this is complicated by the nonlinearity of sgn, and by the fact that only eigenvectors with positive eigenvalues are fixed.

Unlike chapter 5, we have never assumed in this chapter that the matrix T was symmetric. This makes Theorem 6.4 very general indeed.

The formulation of Theorem 6.4 is subtle, the condition *for every set* being very important. As we saw on page 13, one particular $n \times n$ matrix T can have more than n fixed vectors, and they are not spurious patterns, because we did not consider learning.

6.3 Problems

1. What boundary conditions does the equation

$$C^m_{n+1} = C^{m-1}_n + C^{m-1}_{n+1} \qquad (6.18)$$

need?

2. Describe how a dynamic, fully interconnected, neural network recognizes patterns. Do this without giving practical examples. Then, in a second part of your answer, show how some practical problems can be reduced to pattern recognition (you will of course give examples in this part).

3. Are the weights in a network derived from a biological concept?

4. Suppose that a neural network is used to segment a medical image. Assume an idealized case where you always expect to find the same number of segments, and each segment has the same possible classification, e.g., tumor present or no tumor present. To which of the following will the capacity of the network be most relevant: the number of segments, the number of classes per segment, or the product of both?

6.4 Project: a Neural Network Chip

Neural networks are parallel systems, and their full potential can only be realized if the neurons can operate in parallel and if the states of the neurons can be communicated fast to the other neurons they are connected to. This can be realized in a VLSI implementation, a neural network chip.

The currently available neural network chips implement only a few tens of neurons fully in parallel, and the aim of this project is to let you discover why this is the case.

The building blocks available for low-level digital integrated circuit design are inverters, NAND (NOT-AND), and NOR (NOT-OR) gates. They are illustrated in figures 6.6 and 6.7.

They show the location of the features used to make the gates. Different shades represent metal, silicon oxide, and silicon doped with

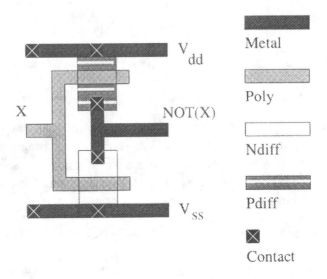

Figure 6.6: A layout of an inverter in CMOS technology.

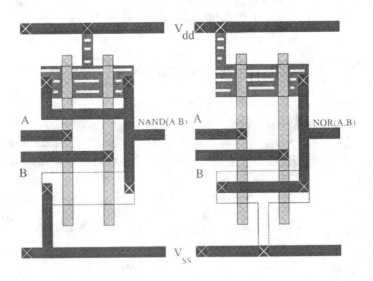

Figure 6.7: Possible layouts for a NAND and a NOR gate.

several impurities. The whole is embedded in a silicon substrate. Together they form the transistors and their connecting wires that make up the gates.

The size of the features determines how many gates will fit on the chip. If you can decompose the neurons in a collection of gates, this will tell you how many neurons you can fit on the chip. The features become smaller as the technology evolves, but it is quite realistic to assume that the width of the lines in figures 6.6 and 6.7 is 0.5 micron. From this, you can estimate the area that the gates occupy on chip.

All we have to do now is to reduce to gates the neural network with dynamics

$$x_i(t+1) = \text{sgn}\left(\sum_{j=1}^{n} T_{ij} x_j(t)\right), \quad i = 1, \ldots, n, \qquad (6.19)$$

and weights

$$T_{ij} = \sum_{\alpha=1}^{m} (x_i^\alpha x_j^\alpha - \delta_{ij}), \quad i = 1, \ldots, n, \quad j = 1, \ldots, n. \qquad (6.20)$$

This means that the threshold functions have to be converted to Boolean functions. Although it could be custom designed, we do not have the function sgn as a ready-made building block. Moreover, if we insisted on using threshold functions, we would have to implement the multiplication of the weights with the states, and the addition of these products. There would be as many adders as neurons on chip, because the neurons have to be able to operate in parallel, and this would take up too much space. The Boolean function will be more economical in size.

Given a threshold function, you find the corresponding Boolean function by first transforming the input from a (0,1) representation to a (-1,+1) representation, applying the threshold function, and then reconverting the output back from (-1,+1) to (0,1). This way you find for example, that the function $\text{sgn} 4x_1 - 7x_2$ is equivalent to $\text{NOT}(X_2)$, where X_2 is the Boolean variable corresponding to x_2.

In order to estimate how many neurons fit on a chip of 1cm^2, we want to find the worst case Boolean function, the function that, when implemented with gates, takes up the largest area. To simplify matters, assume that every neuron will have to implement this same, worst case

function. You can also assume that all gates have the same area, and
that you do not need extra space for wiring them together. However,
if you do have access to a VLSI mask layout editor, produce the actual layout, taking into account the design rules that specify how close
together the transistors and other features can be.

Which threshold function needs the largest number of gates to
implement? You could go through all possible combinations of the
weights, and each time see how many gates the function needs, by syn-
thesizing it using Karnaugh maps or the Quine-McCluskey algorithm
[52]. The following considerations will restrict your search space:

1. if all weights change sign, you obtain the complement of the original Boolean function, requiring the same number of gates plus
   one inverter, not a significant difference,

2. multiplying all weights by the same positive number does not
   change the resulting threshold function or Boolean function,

3. if the sum-of-outerproducts is chosen, the weights take on integer
   values in the range $[-m, +m]$, where m is lower than the capacity
   of the network.

If you have no special exclusive-OR building blocks, then the function exclusive-OR of n variables is the most complex Boolean function
in terms of area. Can you show that a threshold function of n variables
can be equivalent to an exclusive-OR function of n variables?

If at any stage in this project, the circuit becomes too complex, you
can try to extrapolate from simpler circuits. For example, you could
construct the worst case Boolean function for 3, 4, 5, and 6 neurons,
count the number of gates, and extrapolate for more neurons. From
this, could could even estimate an area. However, if you can do some
discrete mathematics, you will be able to find the worst case Boolean
function for an arbitrary number of neurons.

Once you have decided on the most complex Boolean function that
a neuron can have to calculate, count the number of gates, and make
a schema of how you would wire up the gates, based on the layouts
in figures 6.6 and 6.7. Also take into account that every neuron will
have to implement the same function. You have to arrange this layout
so that it fits in a square of 1cm^2. To have a complete layout, you
would also have to wire the outputs of the neurons back to the inputs

to the Boolean functions, and provide a way for storing the values of the neurons. The area taken up by this will be negligible compared to the area of the Boolean function layouts.

You are now in a position to answer the question: how many neurons fit on a chip of 1cm^2? If your application needs more neurons, you will have to reuse neurons, or use additional chips. What happens with the connections between reused neurons, or between neurons on different chips?

Chapter 7

Capacity from a Signal to Noise Ratio

In this chapter, we answer the same question as in the last one, namely how many equilibria or fixed vectors the network has. The difference is that we have made a particular choice for the weight matrix, using the sum-of-outerproducts [60]. Because an extra restriction is imposed on the weight matrix, one expects to find a lower capacity, and we will see that this is indeed the case.

The sum-of-outerproduct rule is a popular choice for the weights, and therefore the calculations in this chapter are of great importance. Moreover, the techniques presented here are used in calculating many other properties of neural networks.

We follow an approach loosely based on information theory. The success of this technique has been illustrated in [80, 16, 23, 76, 68, 90, 100, 66]. There exists a complementary theory based on spin glasses, for which we refer the reader to [43, 10, 103, 24, 64, 82].

7.1 Sum-of-Outerproduct Weights

We will be studying the network

$$x_i(t+1) = \text{sgn}\left(\sum_{j=1}^{n} T_{ij}x_j(t)\right), \quad i = 1, \ldots, n, \tag{7.1}$$

with

$$\text{sgn}x = \begin{cases} 1, & x \geq 0, \\ -1, & x < 0. \end{cases}$$

For simplicity, we do not consider thresholds. They can always be introduced by clamping some neurons, see page 34.

We want to store m patterns $\mathbf{x}^\alpha, \alpha = 1, \ldots, m$, so that they are all stable, i.e.

$$x_i^\alpha = \text{sgn}\left(\sum_{j=1}^n T_{ij}x_j^\alpha\right), \quad i = 1, \ldots, n, \quad \alpha = 1, \ldots, m. \quad (7.2)$$

For the synapses, we will choose the sum-of-outerproducts

$$T_{ij} = \frac{1}{n}\sum_{\alpha=1}^m (x_i^\alpha x_j^\alpha - \delta_{ij}), \quad i = 1, \ldots, n, \quad j = 1, \ldots, n, \quad (7.3)$$

with

$$\delta_{ij} = \begin{cases} 0, & i \neq j, \\ 1, & i = j. \end{cases} \quad (7.4)$$

Remark that the matrix T is symmetric, with zero diagonal. The number $1/n$ is a normalizing factor. As we know from chapter 6 that $m \leq n$, division by n brings all weights in the interval $[-1, +1]$.

Bit i of pattern α will be stable if (7.2) is valid for a fixed i. This will be fulfilled if $\sum_j T_{ij}x_j^\alpha$ has the same sign as x_i^α. In fact, this condition gives stability for more general update rules than (7.1).

Two numbers have the same sign if their product is positive, so that the stability condition for bit i is

$$\left(\sum_{j=1}^n T_{ij}x_j^\alpha\right)x_i^\alpha \geq 0. \quad (7.5)$$

The result we are looking for is the largest number of patterns m as a function of the number of neurons n, such that condition (7.5) holds for all n bits. Indeed, condition (7.5) implicitly defines m as a function of n. The rest of this section will consist of an elaboration of (7.5). The calculation is a bit long winded, but not difficult.

First, substitute (7.3) in (7.5), and do some reordering.

$$\frac{1}{n}\sum_{j=1}^{n}\left(\sum_{\beta=1}^{m}x_i^{\beta}x_j^{\beta}-\delta_{ij}\right)x_j^{\alpha}x_i^{\alpha}\geq 0 \iff \tag{7.6}$$

$$\frac{1}{n}\sum_{j\neq i}^{n}\sum_{\beta=1}^{m}x_i^{\beta}x_j^{\beta}x_j^{\alpha}x_i^{\alpha}\geq 0 \iff$$

$$\left(\frac{1}{n}\sum_{j\neq i}^{n}x_i^{\alpha}x_j^{\alpha}x_j^{\alpha}+\frac{1}{n}\sum_{j\neq i}^{n}\sum_{\beta\neq\alpha}^{m}x_i^{\beta}x_j^{\beta}x_j^{\alpha}\right)x_i^{\alpha}\geq 0 \iff$$

$$\left(\frac{n-1}{n}x_i^{\alpha}+\frac{1}{n}\sum_{j\neq i}^{n}\sum_{\beta\neq\alpha}^{m}x_i^{\beta}x_j^{\beta}x_j^{\alpha}\right)x_i^{\alpha}\geq 0. \tag{7.7}$$

Suppose now that the patterns one wants to store are random. We assume this in order to keep the calculation simple. For correlated patterns, the calculations are similar, but more complicated. The bit x_i^{α} now becomes a random variable ξ_i^{α}, uniformly distributed,

$$\Pr(\xi_i^{\alpha}=-1)=\Pr(\xi_i^{\alpha}=+1)=\frac{1}{2}. \tag{7.8}$$

As we are working with random variables now, it is not possible any more to tell whether (7.7) holds or not, but we can calculate the probability that the inequality holds. We will want this probability to be one.

The product term in (7.7) now contains the $(n-1)(m-1)$ random variables $\xi_i^{\beta}\xi_j^{\beta}\xi_i^{\alpha}\xi_j^{\alpha}, j=1,\ldots,n, j\neq i, \beta=1,\ldots,m, \beta\neq\alpha$. They are not independent, but we will suppose they are. For the calculations without this assumption, see [80].

For the random variable ξ_i^{α}, the mean is $-1\frac{1}{2}+1\frac{1}{2}=0$, and the variance $(-1-0)^2\frac{1}{2}+(1-0)^2\frac{1}{2}=1$. You can verify that the same holds for the product of four such variables. The central limit theorem [91] now states that for $(n-1)(m-1)\to\infty$,

$$\sum_{j\neq i}^{n}\sum_{\beta\neq\alpha}^{m}\xi_i^{\beta}\xi_j^{\beta}\xi_i^{\alpha}\xi_j^{\alpha} \tag{7.9}$$

is a Gaussian random variable with mean 0 and variance $(n-1)(m-1)$, the sum of the individual variances. A simple transformation tells us that

$$\frac{1}{n} \sum_{j \neq i}^{n} \sum_{\beta \neq \alpha}^{m} \xi_i^\beta \xi_j^\beta \xi_i^\alpha \xi_j^\alpha \tag{7.10}$$

is a Gaussian random variable, too, with mean 0 and variance

$$\sigma^2 = \frac{(n-1)(m-1)}{n^2} \tag{7.11}$$

Consider the first factor in the product (7.7), with random variables.

$$\frac{n-1}{n} \xi_i^\alpha + \frac{1}{n} \sum_{j \neq i}^{n} \sum_{\beta \neq \alpha}^{m} \xi_i^\beta \xi_j^\beta \xi_j^\alpha \tag{7.12}$$

The first term is a signal term, and the second term is a noise term, describing how the patterns $\beta = 1, \ldots, m, \beta \neq \alpha$ interfere with the stability of the pattern α. The division of a sum (7.6) in a signal and noise term is a well known technique from signal processing. We use it here to calculate the capacity of a neural network.

The condition for applying the central limit theorem was $(n-1)(m-1) \to \infty$. In addition, we will assume here that $n \to \infty$. This does not follow automatically, because we do not know at this stage what sort of function m is of n. For example, it could be that $m = 1/n^2$.

Rewriting condition (7.7) in terms of random variables, we obtain

$$\lim_{n \to \infty, (n-1)(m-1) \to \infty} \Pr\left[\left(\frac{n-1}{n} \xi_i^\alpha + \frac{1}{n} \sum_{j \neq i}^{n} \sum_{\beta \neq \alpha}^{m} \xi_i^\beta \xi_j^\beta \xi_j^\alpha \right) \xi_i^\alpha \geq 0 \right] = 1 \Leftrightarrow$$

$$\lim_{n \to \infty, (n-1)(m-1) \to \infty} \Pr\left(\frac{1}{n} \sum_{j \neq i}^{n} \sum_{\beta \neq \alpha}^{m} \xi_i^\beta \xi_j^\beta \xi_i^\alpha \xi_j^\alpha \geq -1 \right) = 1 \Longleftrightarrow \tag{7.13}$$

$$\lim_{n \to \infty, (n-1)(m-1) \to \infty} \frac{1}{\sqrt{2\pi\sigma^2}} \int_{-1}^{\infty} e^{-\frac{x^2}{2\sigma^2}} \, dx = 1. \tag{7.14}$$

The integral is visualized in figure 7.1 The limit will be one if the variance of the probability distribution becomes very small, as indicated in figure 7.2.

The stability condition (7.14) has to hold for n independent bits, so that the condition for stability of the complete pattern α is

$$\lim_{n \to \infty, (n-1)(m-1) \to \infty} \left(\frac{1}{\sqrt{2\pi\sigma^2}} \int_{-1}^{\infty} e^{-\frac{x^2}{2\sigma^2}} \, dx \right)^n = 1. \tag{7.15}$$

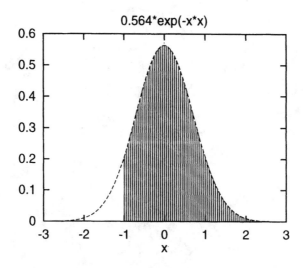

Figure 7.1: The shaded region is the probability that $x \geq -1$.

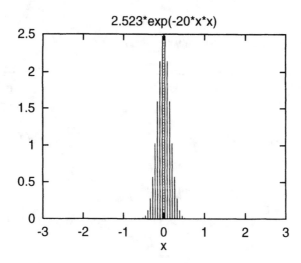

Figure 7.2: For small σ, $\Pr(x \geq -1)$ nearly equals 1.

We want to prove now that

$$\lim_{n\to\infty,(n-1)(m-1)\to\infty} \sigma^2 = 0.$$

This does not follow trivially from (7.11), because we do not know how m varies with n. We will assume

$$\lim_{n\to\infty,(n-1)(m-1)\to\infty} \sigma^2 = c > 0,$$

and try to derive a contradiction.

If we know that the limit of the variance is c, we can use this in (7.15), which becomes

$$\lim_{n\to\infty} \left(\frac{1}{\sqrt{2\pi c}} \int_{-1}^{\infty} e^{-\frac{x^2}{2c}} \, dx \right)^n = 1. \tag{7.16}$$

This is impossible, because

$$\frac{1}{\sqrt{2\pi c}} \int_{-1}^{\infty} e^{-\frac{x^2}{2c}} \, dx < 1.$$

As a result, we have established

$$\lim_{n\to\infty,(n-1)(m-1)\to\infty} \sigma^2 = 0.$$

We will now use an asymptotic approximation to the Gaussian integral for small σ. The relevant formulas can be found in [4], and they are

$$\frac{1}{\sigma\sqrt{2\pi}} \int_{-\infty}^{x} e^{-\frac{(t-m)^2}{2\sigma^2}} \, dt = \frac{1}{2}\left(1 + \text{erf}\left(\frac{x-m}{\sigma\sqrt{2}} \right) \right),$$

$$\text{erf}(-x) = -\text{erf}x,$$

$$\sqrt{\pi}x e^{x^2}(1 - \text{erf}x) \approx 1, \quad x \to \infty.$$

Remark that if σ is small, $1/\sigma$ is large. Condition (7.15) is equivalent to

$$\lim_{n\to\infty,(n-1)(m-1)\to\infty} \left(1 - \frac{1}{\sqrt{2\pi\sigma^2}} \int_{-\infty}^{-1} e^{-\frac{x^2}{2\sigma^2}} \, dx \right)^n = 1 \iff \tag{7.17}$$

$$\lim_{n\to\infty,(n-1)(m-1)\to\infty} \left(1 - \frac{1}{2} - \frac{1}{2}\text{erf}\left(-\frac{1}{\sqrt{2}\sigma} \right) \right)^n = 1 \iff$$

$$\lim_{n\to\infty,(n-1)(m-1)\to\infty} \left(\frac{1}{2}\left[1 - \mathrm{erf}\left(-\frac{1}{\sqrt{2}\sigma}\right)\right]\right)^n = 1 \iff$$

$$\lim_{n\to\infty,(n-1)(m-1)\to\infty} \left(\frac{1}{2}\left[1 + \mathrm{erf}\left(\frac{1}{\sqrt{2}\sigma}\right)\right]\right)^n = 1 \iff$$

$$\lim_{n\to\infty,(n-1)(m-1)\to\infty} \left(\frac{1}{2}\left[1 + 1 - \frac{\sqrt{2}\sigma}{\sqrt{\pi}}e^{-\frac{1}{2\sigma^2}}\right]\right)^n = 1 \iff (7.18)$$

$$\lim_{n\to\infty,(n-1)(m-1)\to\infty} \left[1 - \frac{n\sqrt{2}\sigma}{2\sqrt{\pi}}e^{-\frac{1}{2\sigma^2}}\right] = 1 \iff (7.19)$$

$$\lim_{n\to\infty,(n-1)(m-1)\to\infty} n\sigma e^{-\frac{1}{2\sigma^2}} = 0 \iff (7.20)$$

$$\lim_{n\to\infty,(n-1)(m-1)\to\infty} n\sqrt{\frac{m-1}{n}}e^{-\frac{n}{2(m-1)}} = 0. \qquad (7.21)$$

This last equation now defines m implicitly as a function of n. There is no straightforward way to solve for m, and we will make some guesses, and check whether the limit is zero.

First guess $m - 1 = n$. We find

$$\lim_{n\to\infty} ne^{-\frac{1}{2}} = \infty, \qquad (7.22)$$

so we reduce m for our second guess, $m - 1 = n/\ln n$. We obtain

$$\lim_{n\to\infty} n\frac{1}{\sqrt{\ln n}}e^{-\frac{\ln n}{2}} = \lim_{n\to\infty} \frac{n}{\sqrt{n\ln n}} = \lim_{n\to\infty} \sqrt{\frac{n}{\ln n}} = \infty. \qquad (7.23)$$

For our third guess, we reduce m even more, to $m - 1 = n/(2\ln n)$. The limit is now

$$\lim_{n\to\infty} n\frac{1}{\sqrt{2\ln n}}e^{-\ln n} = \lim_{n\to\infty} \frac{1}{\sqrt{2\ln n}} = 0, \qquad (7.24)$$

and we have finally found a valid value for m. Remark that for $m - 1 = n/(2\ln n)$, $n \to \infty$ implies $(n-1)(m-1) \to \infty$.

It may be possible that a value for $m - 1$ between $n/\ln n$ and $n/(2\ln n)$ gives a zero limit, but we will be content here with the solution $m - 1 = n/(2\ln n)$.

As a conclusion, we have proven the following

Theorem 7.1 *A pattern will be stable with probability 1 for a Hopfield network with sum-of-outerproduct weights, if $n \to \infty$ and the number* *of patterns m obeys the condition*

$$m - 1 \leq \frac{n}{2\ln n}. \tag{7.25}$$

7.2 Capacity Dependent on Representation

Up to now, we have always been using +1 and -1 for the states of the neurons. This is not common, in digital electronics one usually works with the binary 0 and 1. One can of course always transform a pattern $\mathbf{v}^\alpha \in \{0,1\}^n$ to a pattern $\mathbf{x}^\alpha \in \{-1,+1\}^n$ via the transformation

$$2v_i^\alpha - 1 = x_i^\alpha, i = 1, \ldots, n,$$

but there is much more to that. We use here a derivation from [19].

In the following, $\mathbf{v}$ will always be used for a pattern with bits with values 0 and 1, and $\mathbf{x}$ for a pattern with +1 and -1 bits. With the $\mathbf{v}$-notation, the sum-of-outerproducts rule (7.3) becomes

$$T_{ij} = \frac{1}{n} \sum_{\alpha=1}^{m} \left((2v_i^\alpha - 1)(2v_j^\alpha - 1) - \delta_{ij} \right), \quad i = 1, \ldots, n, \quad j = 1, \ldots, n. \tag{7.26}$$

The condition (7.5) for a stable bit i becomes

$$\left(\sum_{j=1}^{n} T_{ij}(2v_j^\alpha - 1) \right) (2v_i^\alpha - 1) \geq 0. \tag{7.27}$$

We will change the input to neuron i to

$$\sum_{j=1}^{n} T_{ij} \frac{1}{2}[(1 - \lambda) + (1 + \lambda)(2v_j^\alpha - 1)], \quad 0 \leq \lambda \leq 1.$$

When $\lambda = 0$, we find that the input is just $\sum T_{ij}v_j^\alpha$, the binary case. When $\lambda = 1$, the input is $\sum T_{ij}(2v_j^\alpha - 1) = \sum T_{ij}x_j^\alpha$, the +1,-1 case, also called the bipolar case.

The new output is still between -1 and 1, so the condition for a stable bit i is

$$\left(\sum_{j=1}^{n} T_{ij} \frac{1}{2}[(1 - \lambda) + (1 + \lambda)(2v_j^\alpha - 1)] \right) (2v_i^\alpha - 1) \geq 0. \tag{7.28}$$

We choose to keep the weights (7.26) in the same form.

Just as in the preceding section, we switch from v_i^α to uniformly distributed random bits γ_i^α. We can now substitute the weights, and split into a signal and noise term, taking into account that $(2\gamma_i^\beta - 1)(2\gamma_i^\beta - 1) - \delta_{ij} = 0$.

$$\Pr \left\{ \left[\sum_{j=1}^{n} \frac{1}{n} \sum_{\beta=1}^{m} \left((2\gamma_i^\beta - 1)(2\gamma_j^\beta - 1) - \delta_{ij} \right) \right. \right.$$
$$\left. \left. \times \frac{1}{2}[(1 - \lambda) + (1 + \lambda)(2\gamma_j^\alpha - 1)] \right] (2\gamma_i^\alpha - 1) \geq 0 \right\} = 1 \Leftrightarrow \quad (7.29)$$

$$\Pr \left\{ \left[\frac{1}{2n} \sum_{j \neq i}^{n} (2\gamma_i^\alpha - 1)(2\gamma_i^\alpha - 1)(2\gamma_j^\alpha - 1)[(1 - \lambda) + (1 + \lambda)(2\gamma_j^\alpha - 1)] \right. \right.$$
$$+ \frac{1}{2n} \sum_{j \neq i}^{n} \sum_{\beta \neq \alpha}^{m} (2\gamma_i^\alpha - 1)(2\gamma_i^\beta - 1)(2\gamma_j^\beta - 1)$$
$$\left. \left. \times [(1 - \lambda) + (1 + \lambda)(2\gamma_j^\alpha - 1)] \right] \geq 0 \right\} = 1 \Leftrightarrow \quad (7.30)$$

$$\Pr \left\{ \left[\frac{1}{2n} \sum_{j \neq i}^{n} [(1 + \lambda) + (1 - \lambda)(2\gamma_j^\alpha - 1)] \right. \right.$$
$$+ \frac{1}{2n} \sum_{j \neq i}^{n} \sum_{\beta \neq \alpha}^{m} (2\gamma_i^\alpha - 1)(2\gamma_i^\beta - 1)(2\gamma_j^\beta - 1)$$
$$\left. \left. \times [(1 - \lambda) + (1 + \lambda)(2\gamma_j^\alpha - 1)] \right] \geq 0 \right\} = 1. \quad (7.31)$$

This is again a sum of a signal term and a noise term. Let's first analyze the signal term

$$\frac{1}{2n} \sum_{j \neq i}^{n} [(1 + \lambda) + (1 - \lambda)(2\gamma_j^\alpha - 1)].$$

The random variable $2\gamma_j^\alpha - 1$ has mean 0, so that $(1+\lambda)+(1-\lambda)(2\gamma_j^\alpha - 1)$ has mean $1 + \lambda$. The variance

$$\sum_{\gamma_j^\alpha = 0}^{+1} [(1 - \lambda)(2\gamma_j^\alpha - 1)]^2 \Pr(\gamma_j^\alpha) = (1 - \lambda)^2 \quad (7.32)$$

From this result follows that $\frac{1}{2n}[(1 + \lambda) + (1 - \lambda)(2\gamma_j^\alpha - 1)]$ has mean $(1 + \lambda)/2n$ and variance $(1 - \lambda)^2/4n^2$.

We can now use the central limit theorem to find that the signal term

$$\frac{1}{2n} \sum_{j \neq i}^{n} [(1 + \lambda) + (1 - \lambda)(2\gamma_j^\alpha - 1)] \tag{7.33}$$

has a Gaussian distribution with mean

$$\mu = \frac{n-1}{2n}(1 + \lambda), \tag{7.34}$$

and variance

$$\frac{n-1}{4n^2}(1 - \lambda)^2. \tag{7.35}$$

The noise term

$$\frac{1}{2n} \sum_{j \neq i}^{n} \sum_{\beta \neq \alpha}^{m} (2\gamma_i^\alpha - 1)(2\gamma_i^\beta - 1)(2\gamma_j^\beta - 1)[(1 - \lambda) + (1 + \lambda)(2\gamma_j^\alpha - 1)]$$

is a sum of $(n-1)(m-1)$ random variables

$$\frac{1}{2n}(2\gamma_i^\alpha - 1)(2\gamma_i^\beta - 1)(2\gamma_j^\beta - 1)[(1 - \lambda) + (1 + \lambda)(2\gamma_j^\alpha - 1)]. \tag{7.36}$$

They are not independent, but as in the preceding section, we will suppose that they are. The random variables (7.36) have mean 0, and variance

$$\frac{1}{4n^2} \left[\frac{1}{2}[(1 - \lambda) + (1 + \lambda)]^2 + \frac{1}{2}[(1 - \lambda) - (1 + \lambda)]^2 \right] = \frac{1 + \lambda^2}{2n^2}. \tag{7.37}$$

Using the central limit theorem, we find that the noise term has a Gaussian distribution with mean 0 and variance

$$\sigma^2 = \frac{(n-1)(m-1)}{2n^2}(1 + \lambda^2). \tag{7.38}$$

Condition (7.31) is very similar to (7.13). The signal term now has a variance (7.35), but this variance becomes zero in the limit $n \to \infty$. If we assume the limit has been taken, so that the variance is zero, condition (7.31) contains the area under a Gaussian probability distribution, as illustrated in figure 7.1, but now with $-\mu$ instead of -1 for the lower boundary.

If we now also take into account that n independent bits have to be stable, condition (7.35) has become equivalent to (remark the similarity with (7.15))

$$\lim_{n\to\infty,(n-1)(m-1)\to\infty} \left(\frac{1}{\sqrt{2\pi\sigma^2}}\int_{-\mu}^{\infty} e^{-\frac{x^2}{2\sigma^2}}\,dx\right)^n = 1 \iff (7.39)$$

$$\lim_{n\to\infty,(n-1)(m-1)\to\infty} \left(1 - \frac{1}{\sqrt{2\pi\sigma^2}}\int_{-\infty}^{-\mu} e^{-\frac{x^2}{2\sigma^2}}\,dx\right)^n = 1 \iff$$

$$\lim_{n\to\infty,(n-1)(m-1)\to\infty} \left(1 - \frac{1}{2} - \frac{1}{2}\mathrm{erf}\left(-\mu\sqrt{\frac{1}{2\sigma^2}}\right)\right)^n = 1 \iff$$

$$\lim_{n\to\infty,(n-1)(m-1)\to\infty} \left\{\frac{1}{2}\left[1 + \mathrm{erf}\left(\frac{\mu}{\sqrt{2}\sigma}\right)\right]\right\}^n = 1. \qquad (7.40)$$

The argument of the error function is

$$\frac{\mu}{\sqrt{2}\sigma} = \frac{n-1}{2n}(1+\lambda)\sqrt{\frac{2n^2}{(1+\lambda^2)(n-1)(m-1)}}\frac{1}{\sqrt{2}}$$

$$= \frac{1}{2}\frac{1+\lambda}{\sqrt{1+\lambda^2}}\sqrt{\frac{n-1}{m-1}}, \qquad (7.41)$$

and in order to simplify calculations we will suppose that this argument is large as $n \to \infty$, i.e. m increases slower than n.

Condition (7.40) is now

$$\lim_{n\to\infty,(n-1)(m-1)\to\infty} \left\{\frac{1}{2}\left[2 - \frac{1}{\sqrt{\pi}}\frac{\sigma\sqrt{2}}{\mu}e^{-\frac{\mu^2}{2\sigma^2}}\right]\right\}^n = 1 \iff (7.42)$$

$$\lim_{n\to\infty,(n-1)(m-1)\to\infty} n\frac{\sigma}{\mu}e^{-\frac{\mu^2}{2\sigma^2}} = 0 \iff$$

$$\lim_{n\to\infty,(n-1)(m-1)\to\infty} n\frac{\sqrt{1+\lambda^2}}{1+\lambda}\sqrt{\frac{m-1}{n}}e^{-\frac{1}{4}\frac{(1+\lambda)^2}{1+\lambda^2}\frac{n-1}{m-1}} = 0. \qquad (7.43)$$

We now have again to guess the solution of this equation for m as a function of n. We will immediately give the right guess, but the reader should do some experiments on his or her own. For the guess

$$m - 1 = \frac{n}{\ln n}\frac{(1+\lambda)^2}{4(1+\lambda^2)}$$

the limit (7.43) becomes

$$\lim_{n \to \infty} \frac{n}{2\sqrt{\ln n}} e^{-\ln n} = \lim_{n \to \infty} \frac{1}{2\sqrt{\ln n}} = 0. \qquad (7.44)$$

We can now formulate this result as a theorem.

Theorem 7.2 *A pattern will be stable with probability 1 for the Hopfield network with sum-of-outerproduct weights, and with the output of the neurons*

$$\frac{1}{2}[(1 - \lambda) + (1 + \lambda)x_j], \quad 0 \le \lambda \le 1,$$

if $n \to \infty$ and the number of patterns m obeys the condition

$$m - 1 \le \frac{n}{\ln n} \frac{(1 + \lambda)^2}{4(1 + \lambda^2)}. \qquad (7.45)$$

For $\lambda = 1$, we find

$$m - 1 \le \frac{n}{2\ln n},$$

which is the result (7.25) derived for neurons with states -1 and 1. If we substitute $\lambda = 0$, we find the capacity for neurons with binary states 0 and 1,

$$m - 1 \le \frac{n}{4\ln n},$$

a surprising result, as it shows that the capacity for neurons with binary states is lower that the capacity for neurons with bipolar states.

7.3   Problems

1. How can the capacity results for neural networks be used in the design of a neural network?

2. Is it a contradiction that there is a capacity result n for neural networks, and another capacity result $\frac{n}{2\ln n}$?

3. Describe the main capacity results in neural networks.

4. Do the patterns need to have statistically independent bits in the derivation of the $n/(2\ln n)$ capacity result?

5. Do the patterns you store in a neural network have to be uncorrelated?

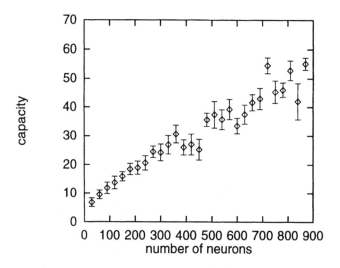

Figure 7.3: Number of stable patterns versus number of neurons. The sum-of-outerproducts rule was used, and the standard deviation was estimated over 30 trials.

7.4 Project: Capacity and Attractor Size

Program a fully interconnected neural network. Provide simple graphic output, but also output in a file that you can analyze later with another program.

Program the sum-of-outerproducts rule. Program the generation of random patterns. Structure your software carefully. It will be the experimental device upon which you will rely.

Fix the size of the network. Store a set of patterns. Verify for each pattern if it is stable, and examine how less and less patterns remain stable as you store more and more patterns. If you call pattern α stable if all the neurons are stable,

$$x_i^\alpha = \text{sgn}\left(\sum_{j=1}^n T_{ij}x_j^\alpha\right), \quad i = 1,\ldots,n, \tag{7.46}$$

you will find for the capacity a relation as in figure 7.3. For what size of the network can you draw general conclusions?

You could define a kind of approximate stability, where all neurons are stable, except for a fraction s. Draw several capacity curves, for

s decreasing from 0.1 to zero. What happens for a large number of neurons?

The sum-of-outerproducts rule implicitly makes the diagonal of the weight matrix zero. This is not absolutely necessary [64, 33]. To increase the stability, would you make the diagonal elements positive or negative? Remember the remarks made on page 122. Calculate the new capacity, with the non-zero diagonal.

Also test for spurious patterns (fundamental memories that are not stored patterns). How many are there? How far are they from the fundamental patterns? Does the size of the diagonal have an effect? This requires several exhaustive searches. You will have to program them efficiently if you want to examine any network of a reasonable size.

Return now to the sum-of-outerproducts rule, with all bits stable in the fundamental memories, and zero diagonal for the weight matrix. The attraction regions around the fundamental memories are in general not balls in the n-dimensional space. A ball in n-dimensional space of radius r around a fundamental memory is the set of all points that are Hamming distance r or less away of the fundamental memory.

You can measure the deformation of the attraction region around a fundamental memory by considering the ratio of the radius of the largest ball that contains it, to the radius of the smallest ball that fits in it. Elongated attraction regions will have a large ratio. Better ways of describing the attraction regions are possible [18]. Are the attraction regions deformed as you store more and more patterns in a network?

Now, abandon the sum-of-outerproducts rule, make a random choice for the weight matrix T, but keep the diagonal zero. Does the capacity (number of stable patterns) increase?

What happens to the conclusions you have drawn when the patterns to be stored are correlated?

You have been experimenting with a general kind of neural networks. Formulate some design rules that could help a person that has some patterns to store and retrieve associatively and who has to choose a particular architecture.

Chapter 8

Neural Networks and Markov Chains

The networks we have studied up to now were all operating in a deterministic fashion. If the network was in a particular state, the next state was always the same. This seems the very model of a reliable system that every engineer wants. If the network is used for optimization however, e.g. to solve the Travelling Salesman Problem [99], a network operating deterministically will usually not find the best solution. When converging towards an equilibrium, it will get stuck into a local optimum.

To move out of this local optimum towards the global optimum, one can add noise to the operation of the system. This is the principle of simulated annealing [1]. We will show in this chapter that a neural networks offers in a very natural way a parallel architecture for simulated annealing. The neural network operating with noise is also called a Boltzmann machine [25].

For the analysis of the network, we follow [25, 29, 51]. We will use some elementary notions from Markov chains [26].

8.1 Simulated Annealing

In this Chapter, we will consider a network of n neurons, whose states are random variables $\xi_i, i = 1, \ldots, n$. The random variables can take on the values +1 or -1. Note that this is different from last chapter, where the patterns to be stored were random. Here, it is the states of the neurons themselves that are random, as if the neurons were throwing biased dice. This is a convenient way to model uncertainty in our knowledge about the neurons.

In chapters 5, 6, and 7, the discretized time was advancing by one unit at a time, but here we will assume that the time changes in steps of $\tau \approx 0.5$ msec. The neurons update their states synchronously at times

$$0, \tau, 2\tau, \ldots.$$

The signal sent from neuron j to neuron i at time t is

$$T_{ij}\xi_j(t).$$

After initial excitation, the setting of the states of all neurons at time zero, the network operates autonomously, without external input.

We will specify now the update or firing rule. Each neuron calculates

$$F_i(t) = \sum_{j=1}^{n} T_{ij}\xi_j(t) - \theta_i, \tag{8.1}$$

where θ_i is a threshold. This is all very similar to what we have seen before.

In calculating the next state of neuron i, at time $t + \tau$, we do not have a deterministic formula, but only a probability.

$$\Pr(\xi_i(t+\tau) = +1) = \frac{1}{1 + e^{-\beta_i F_i(t)}}, \quad \beta_i \geq 0,$$

$$\Pr(\xi_i(t+\tau) = -1) = \frac{1}{1 + e^{\beta_i F_i(t)}}. \tag{8.2}$$

It can be verified that

$$\Pr(\xi_i(t+\tau) = +1) + \Pr(\xi_i(t+\tau) = -1) = \frac{2 + e^{\beta_i F_i} + e^{-\beta_i F_i}}{(1 + e^{-\beta_i F_i})(1 + e^{\beta_i F_i})} = 1. \tag{8.3}$$

In order to understand the meaning of the number β_i, consider the formula (8.2) in the limit $\beta_i \to \infty$.

$$\Pr(\xi_i(t+\tau) = +1) = \begin{cases} 1 & F_i(t) > 0, \\ 0 & F_i(t) < 0, \\ \frac{1}{2} & F_i(t) = 0, \end{cases}$$

$$\Pr(\xi_i(t+\tau) = -1) \;=\; \begin{cases} 0 & F_i(t) > 0, \\ 1 & F_i(t) < 0, \\ \frac{1}{2} & F_i(t) = 0. \end{cases} \qquad (8.4)$$

For $F_i(t) \neq 0$, this is the deterministic network (5.10).

It is possible to avoid the situation $F_i(t) = 0$ by an adjustment of the thresholds θ_i, as was explained in Theorem 5.1.

Consider now (8.2) in the limit $\beta_i \to 0$.

$$\begin{aligned} \Pr(\xi_i(t+\tau) = +1) &= \frac{1}{2}, \\ \Pr(\xi_i(t+\tau) = -1) &= \frac{1}{2}. \end{aligned} \qquad (8.5)$$

This shows that for $\beta_i \to 0$, the neurons are throwing unbiased dice, they are not influenced by their input F_i any more. We could call this chaos, but it is different from the chaos that can occur in the networks studied in chapter 4.

We will now try to relate β_i to the evolution of the energy. A deterministic equivalent of (8.2) would be

$$x_i(t+\tau) = \operatorname{sgn}\left(\sum_{j=1}^{n} T_{ij}x_j(t) - \theta_i\right). \qquad (8.6)$$

A candidate for the energy function is

$$E = \left(-\sum_{j=1}^{n}\sum_{k=1}^{n} T_{jk}x_j(t)x_k(t)\right) + 2\sum_{j=1}^{n} x_j(t)\theta_j. \qquad (8.7)$$

As usual with an energy function, we will assume that the synapse matrix T is symmetric and has zero diagonal

Calculate now the partial derivative

$$\begin{aligned} \frac{\partial E}{\partial x_i} &= -\sum_{k=1}^{n} T_{ik}x_k - \sum_{j=1}^{n} T_{ji}x_j + 2\theta_i, \\ &= -2\left(\sum_{j=1}^{n} T_{ij}x_j - \theta_i\right). \end{aligned} \qquad (8.8)$$

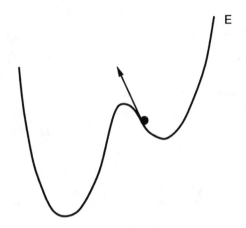

Figure 8.1: The energy can increase during simulated annealing.

Switching to finite differences, this formula becomes

$$\Delta E = -2\Delta x_i \left(\sum_{j=1}^{n} T_{ij} x_j - \theta_i \right) \tag{8.9}$$

One can verify now that, if $\sum_j T_{ij} x_j - \theta_i \geq 0$, then $\Delta x_i \geq 0$, because of (8.6), and $\Delta E \leq 0$. Similarly, if $\sum_j T_{ij} x_j - \theta_i < 0$, then $\Delta x_i \leq 0$, and again $\Delta E \leq 0$, proving that the energy E decreases during the operation of the network.

In the probabilistic situation, however, with the rule (8.2), the condition $\sum_j T_{ij} x_j - \theta_i \geq 0$ can result in $x_i(t+\tau) = -1$ so that, if $x_i(t) = 1$, $\Delta x_i \leq 0$, and $\Delta E \geq 0$. We see here that *the energy can increase if the network operates probabilistically!* This is illustrated in figure 8.1.

How likely is the increase in energy? Without doing complicated calculations, one sees that the energy increases if (8.2) does not obey the sgn rule. This is most likely for small β_i.

So far, we have seen that the probabilistically operating neuron (8.2) becomes deterministic for $\beta_i \to \infty$. As β_i becomes smaller and smaller, it becomes more and more likely that the energy will increase. For $\beta_i = 0$, the neuron i behaves like an unbiased dice. This is true for every neuron.

This leads to the interpretation of $1/\beta_i$ as a temperature. For low temperature, the system is deterministic, for high temperature, the system is noisy, and chaotic.

We have proven

Theorem 8.1 *In a probabilistic network, the energy can increase if the temperature is not zero.*

This interpretation is extremely important for applications where a neural network is used such that its energy function corresponds to a cost function. By letting the network evolve in time, the cost function is minimized. Usually, one finds a local minimum. To find the global minimum, the temperature is increased, so that the energy can increase. After a while, the network is in another region of the state space, and the temperature can be decreased again. By timing this process carefully [1], one will find the global optimum.

The state of the network is a vector $\boldsymbol{\xi}$ in the state space $\{-1, +1\}^n$. From (8.2), we can calculate the probability that neuron i has a certain value, given the state of the network one time step earlier,

$$\Pr(\xi_i^\alpha(t + \tau) = x_i^\alpha | \boldsymbol{\xi}^\beta(t) = \mathbf{x}^\beta) =$$

$$\left\{ 1 + \exp\left[-\beta_i x_i^\alpha(t) \left(\sum_{j=1}^n T_{ij} x_j^\beta(t) - \theta_i \right) \right] \right\}^{-1}, \qquad (8.10)$$

where α and β are subscripts running over all the states, $\alpha, \beta = 1, \ldots, 2^n$.

The probabilities (8.10) are uncorrelated for different i, they depend only on the state one time step earlier. From them, we can calculate the transition rate per unit time, for a transition from state β to a state α in the state space,

$$
\begin{aligned}
R_{\alpha\beta} &= \frac{1}{\tau} Q_{\alpha\beta} = \frac{1}{\tau} \Pr(\boldsymbol{\xi}^\alpha(t + \tau) = \mathbf{x}^\alpha | \boldsymbol{\xi}^\beta(t) = \mathbf{x}^\beta) \\
&= \frac{1}{\tau} \prod_{i=1}^n \left\{ 1 + \exp\left[-\beta_i x_i^\alpha(t) \left(\sum_{j=1}^n T_{ij} x_j^\beta(t) - \theta_i \right) \right] \right\}^{-1},
\end{aligned}
$$

$$\alpha, \beta = 1, \ldots, 2^n. \qquad (8.11)$$

This shows that the states of a neural network follow a Markov chain. For the theory of Markov chains, see [26].

The transition rates $R_{\alpha\beta} \geq 0$, and even $R_{\alpha\beta} > 0$, if $\beta < \infty$. The numbers $Q_{\alpha\beta}$ are the *transition probabilities*. From now on, we drop the restriction that the synapse matrix T has to be symmetric.

8.2   The Fokker-Planck Equation

In a network that operates probabilistically, it is not possible to sketch a particular trajectory in state space, as the next state is only given in a probabilistic sense, via the transition probabilities $Q_{\alpha\beta}$ in (8.11).

What we can talk about in a deterministic sense is the occupation probability of state α at time $n\tau$, denoted by $p_\alpha(n\tau)$. This quantity evolves in a deterministic way according to the equation

$$p_\alpha(n\tau) = \sum_{\beta=1}^{2^n} Q_{\alpha\beta} p_\beta(n\tau - \tau). \tag{8.12}$$

This can be formulated in words as *the occupation probability of state α is the occupation probability of state β multiplied by the probability of a state transition from β to α, and summed over all states β.*

The matrix Q is called stochastic if its column sums equal 1. This means that the sum of the probabilities to go from state β to any other state is 1.

Theorem 8.2 *The matrix Q with elements*

$$Q_{\alpha\beta} = \prod_{i=1}^{n}\left\{1 + \exp\left[-\beta_i x_i^\alpha \left(\sum_{j=1}^{n} T_{ij} x_j^\beta - \theta_i\right)\right]\right\}^{-1},$$
$$\alpha, \beta = 1, \ldots, 2^n \tag{8.13}$$

is stochastic.

Proof.

$$\sum_{\alpha=1}^{2^n} Q_{\alpha\beta} = \sum_{x_1^\alpha=-1}^{1} \sum_{x_2^\alpha=-1}^{1} \cdots \sum_{x_n^\alpha=-1}^{1} \prod_{i=1}^{n} \frac{1}{1 + e^{-\beta_i x_i^\alpha F_i}},$$

$$= \sum_{x_1^\alpha=-1}^{1} \cdots \sum_{x_{n-1}^\alpha=-1}^{1} \prod_{i=1}^{n-1} \frac{1}{1 + e^{-\beta_i x_i^\alpha F_i}} \sum_{x_n^\alpha=-1}^{1} \frac{1}{1 + e^{-\beta_n x_n^\alpha F_n}},$$

$$= \sum_{x_1^\alpha=-1}^{1} \cdots \sum_{x_{n-1}^\alpha=-1}^{1} \prod_{i=1}^{n-1} \frac{1}{1 + e^{-\beta_i x_i^\alpha F_i}},$$

$$= \vdots$$

$$= \sum_{x_1^\alpha = -1}^{1} \frac{1}{1 + e^{-\beta_1 x_1^\alpha F_1}},$$

$$= 1. \tag{8.14}$$

Q.E.D.

We will now try to write down an equation describing how the occupation probabilities evolve in time. Using the fact that Q is a stochastic matrix,

$$p_\alpha(t + \tau) - p_\alpha(t) = \sum_{\beta=1}^{2^n} Q_{\alpha\beta} p_\beta(t) - p_\alpha(t),$$

$$= \sum_{\beta=1}^{2^n} Q_{\alpha\beta} p_\beta(t) - \sum_{\beta=1}^{2^n} Q_{\beta\alpha} p_\alpha(t). \tag{8.15}$$

Dividing the left and right hand side by τ, we find an expression for the finite difference of the occupation probabilities,

$$\frac{p_\alpha(t + \tau) - p_\alpha(t)}{\tau} = \sum_{\beta=1}^{2^n} (R_{\alpha\beta} p_\beta(t) - R_{\beta\alpha} p_\alpha(t)). \tag{8.16}$$

This can be formulated in words as follows: *the change in occupation probability of state α is the probability to be in state β multiplied by the transition rate of going from β to α minus the probability to be in α multiplied by the rate of going from α to β, summed over all states.* This is illustrated in figure 8.2.

In the limit for small τ, equation (8.16) becomes a differential equation, and is called the Fokker-Planck equation, or Master equation, or Chapman-Kolmogorov equation. In this limit, and also for the distinction between the different sorts of equations, there arise many mathematical subtleties, for which we refer the reader to [26].

Finally, we will investigate the *steady-state* solution of (8.16). When the occupation probabilities are not changing anymore in time,

$$\sum_{\beta=1}^{2^n} (R_{\alpha\beta} p_\beta - R_{\beta\alpha} p_\alpha) = 0, \quad \alpha = 1, \ldots, 2^n. \tag{8.17}$$

In this system of linear equations, $R_{\alpha\beta}$ is known, and the p_α are the unknowns. The equations could be solved numerically, if it was not that there are 2^n of them! Instead we will prove

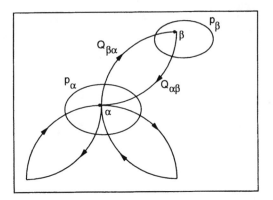

Figure 8.2: A picture of the state space, with states α and β. The probability p_α is represented by a region of size proportional to p_α. Either the probabilities Q or the rates R can be used in the drawing.

Theorem 8.3 *The system of linear equations*

$$\sum_{\beta=1}^{2^n} (R_{\alpha\beta} p_\beta - R_{\beta\alpha} p_\alpha) = 0, \quad \alpha = 1, \ldots, 2^n$$

for the steady state occupation probabilities has a non-zero solution.

Proof. Consider the matrix W with elements

$$W_{\alpha\beta} = R_{\alpha\beta} - \delta_{\alpha\beta} \sum_{\gamma=1}^{2^n} R_{\gamma\alpha}. \tag{8.18}$$

This means

$$W_{\alpha\beta} = \begin{cases} R_{\alpha\beta}, & \alpha \neq \beta, \\ R_{\alpha\alpha} - \sum_{\gamma=1}^{2^n} R_{\gamma\alpha}, & \alpha = \beta, \end{cases}$$

i.e. from each diagonal element of R, the sum of all its column elements has been subtracted.

The system (8.17) is now equivalent to

$$\sum_{\beta=1}^{2^n} W_{\alpha\beta} p_\beta = 0, \quad \alpha = 1, \ldots, 2^n. \tag{8.19}$$

Indeed, (8.19) is equivalent to

$$\sum_{\beta=1}^{2^n} (R_{\alpha\beta}p_\beta - \delta_{\alpha\beta} \sum_{\gamma=1}^{2^n} R_{\gamma\alpha}p_\beta) = 0, \quad \alpha = 1, \ldots, 2^n. \qquad (8.20)$$

The system (8.19) will have a non-zero solution if and only if the determinant $\det W = 0$. This can be verified in the following way.

$$\sum_\alpha W_{\alpha\beta} = \sum_\alpha R_{\alpha\beta} - \sum_\alpha \delta_{\alpha\beta} \sum_\gamma R_{\gamma\alpha} = \sum_\alpha R_{\alpha\beta} - \sum_\gamma R_{\gamma\beta} = 0. \quad (8.21)$$

This proves that the sum of all columns of W is 0, so that its determinant has to be 0.
Q.E.D.

This result is the stepping stone for a further analysis of the steady state occupation probabilities of neural networks, for which we refer to [25].

8.3 Problems

1. How can uncertainty be modelled with neural networks? How do you handle uncertainty in data and their classification? Describe what properties of the network play a role, and also give practical examples.

2. What happens to a neural network at zero temperature? Is Theorem 8.2 still valid?

8.4 Project: Learning with Hidden Units

8.4.1 Deterministic Networks

Sometimes you need to learn a number of fundamental memories that exceeds the capacity of the network that you are using. The solution to this problem is nearly always to increase the number of neurons. But what do these new neurons represent? If, for the fundamental memories, the neurons represent pixels in an image, you could add new neurons representing, for example, a black bar under the image. In this way, however, you have added information to the images that was not originally there, and this will affect the classification.

An alternative is to have the extra neurons represent noise. The problem now becomes how to determine the weights for the connections to and from these extra neurons, because you do not know the desired state of these neurons. As there is randomness involved, the techniques learned in this chapter will help us to solve this problem. At the same time an unexpected connection between deterministic and stochastic networks will arise.

Consider the deterministic network

$$x_i(t+1) = \text{sgn}\left(\sum_{j=1}^{n} T_{ij}x_j(t)\right), \quad i = 1, \ldots, n, \qquad (8.22)$$

with the sum-of-outerproducts

$$T_{ij} = \sum_{\alpha=1}^{m} x_i^\alpha x_j^\alpha, \quad i = 1, \ldots, n, \quad j = 1, \ldots, n. \qquad (8.23)$$

In order to alleviate the notation, we will always assume that the diagonal of the weight matrix is zero, and that $i \neq j$ in expressions such as T_{ij} or $x_i^\alpha x_j^\alpha$. We also assume that m does not exceed the capacity of the network.

Suppose that you did not know that the formula (8.23) makes the m patterns $\mathbf{x}^\alpha, \alpha = 1, \ldots, m$ stable in the network. You would have to guess the weights. You could do this as follows. Start the network in a random state, and observe the product $x_i(t)x_j(t)$ over a period long enough so that the network can settle in an attractor or a limit cycle. Do this several times, more often than there are neurons. Calculate the double average, over time, and over the number of times you restarted the network. Do this for all products, and call these averages

$$< x_i x_j >_f \quad i = 1, \ldots, n, \quad j = 1, \ldots, n. \qquad (8.24)$$

If our choice of weights was very different from the weights (8.23) that store the patterns $\mathbf{x}^\alpha, \alpha = 1, \ldots, m$, then the quantities (8.24) will be very different from the weights (8.23). If, on the other hand, our choice of weights was close to (8.23), the network, when starting from random states, will often be attracted to one of the patterns $\mathbf{x}^\alpha, \alpha = 1, \ldots, m$. If you restart the network often enough, all attractors will appear some of the time. We will call this operation of the network, with a guess for the weights, and random initial states, the *free running* mode of operation.

Now suppose that you again start the network several times, but instead of random initial states you choose the fundamental memories, and you keep the neurons in the state they are for the fundamental memory that you start in. Because the neurons are *clamped*, their states do not change. You can now again observe the products $x_i x_j$. Because the neurons were clamped, we will denote them by

$$< x_i x_j >_c \quad i = 1, \ldots, n, \quad j = 1, \ldots, n. \tag{8.25}$$

If you choose the fundamental memories at random, and you do this often enough, $< x_i x_j >_c$ will be a good approximation to T_{ij}. Indeed, the sum-of-outerproducts (8.23) equals $< x_i x_j >_c$, if every fundamental memory was chosen just once. A uniformly random choice of fundamental memories, and averaging, approximates this.

In this way, the quantity

$$< x_i x_j >_c - < x_i x_j >_f \tag{8.26}$$

tells us how wrong we were with our initial guess for the weights. This leads to the following iterative procedure for adapting the weights in a Hopfield network, starting from an initial guess:

$$\Delta T_{ij} = \eta(< x_i x_j >_c - < x_i x_j >_f), \quad i = 1, \ldots, n, \quad j = 1, \ldots, n, \tag{8.27}$$

where η is a learning rate such as the one we encountered in the back-propagation algorithm (2.5) and (2.6).

Implicit in our reasoning is that all attraction basins have the same size. Otherwise $< x_i x_j >_f$ would approximate $\sum_{\alpha=1}^{m} s_\alpha x_i^\alpha x_j^\alpha$, where s_α is proportional to the size of the attraction basin of $\mathbf{x}^\alpha$. This in itself, however, is a valid alternative to the sum-of-outerproducts rule.

What we have found in (8.27) is an iterative learning rule for a Hopfield network. Because it is iterative, and because random initial states have to be selected repeatedly, it is much slower than the sum-of-outer-products rule. However, it also works when there are hidden neurons, in which case the sum-of-outerproducts rule fails! For a network with hidden neurons, the free running average is the same as defined above. For the clamped average, you now also have to average over random initial states of the hidden neurons.

A good starting point to explore this is to learn the XOR (exclusive-OR) function for two inputs. As we saw on page 45, a feedforward network needs hidden neurons to be able to learn this function. The same is true for a Hopfield network. Assume that the third neuron is the output neuron, then the four patterns to learn are the rows of

$$
\begin{array}{ccc}
+ & + & - \\
- & - & - \\
+ & - & + \\
- & + & +
\end{array}
\tag{8.28}
$$

Try the sum-of-outerproducts rule for a three neuron network, it will not be able to store these patterns. Now add hidden neurons, and train the network with the rule (8.27). After learning, verify if the patterns (8.28) are stable. How many hidden neurons did you need?

8.4.2 Boltzmann Learning

A formula similar to the rule (8.27) exists for stochastic networks, with dynamics (8.2). It is known as Boltzmann learning, because the equilibrium distribution of the stochastic network is a Boltzmann distribution. A formal proof of the Boltzmann learning rule is in [56], but we will extend

$$
\Delta T_{ij} = \eta(< \xi_i \xi_j >_c - < \xi_i \xi_j >_f), \quad i = 1, \ldots, n, \quad j = 1, \ldots, n,
\tag{8.29}
$$

so that the averages over the random variables make sense.

The basic form of (8.27) remains, but we must make clear what is meant by free running and clamped operation of a stochastic network. During operation, neurons are selected at random, and the next state is determined via (8.2). This happens at a certain value for β, which we will assume identical for all neurons.

Because of the non-zero temperature, we cannot expect that the network remains at an attractor, but be want the occupation probabilities of the fundamental memories to be quite high, and those of the other states to be low. This can only happen at fairly low temperature, because at infinite temperature the system is chaotic, with all states equally likely. But at low temperature, it takes a long time for the system to reach is equilibrium distribution. We will therefore use simulated annealing, where the system starts at a high temperature, and the temperature is slowly decreased. Thus, in order to calculate $< \xi_i \xi_j >_f$ or $< \xi_i \xi_j >_c$, we start with a high temperature, let the system approach an equilibrium, decrease the temperature, let the system approach an equilibrium again, etc. At the lowest temperature, we sample the values of the products of the state variables when equilibrium is approached. This gives then the values of $< \xi_i \xi_j >_f$ or $< \xi_i \xi_j >_c$, dependent on whether the system was free running, or whether units were clamped. Finally, apply the rule (8.29) to improve upon your guess of the weights.

You cannot know exactly when the system has reached the equilibrium distribution. Set yourself a range within which the occupation probabilities should not change anymore, to determine when the system is in equilibrium.

Again try to learn the exclusive-OR function. How much longer did it take for the stochastic network to learn, compared to the deterministic network?

The Boltzmann learning rule, derived from theory, has $\eta\beta$ instead of just η in (8.29). The β in this expression is the inverse of the temperature at equilibrium, when the averages have been taken, at the end of the simulated annealing. This value is constant, and can be amalgamated into η.

8.4.3 Mean-field Learning

Waiting for the equilibrium to be reached in a stochastic network takes a long time.

Instead of applying the rules

$$
\begin{aligned}
\Pr(\xi_i(t+1) = +1) &= \frac{1}{1 + e^{-\beta F_i(t)}}, \\
\Pr(\xi_i(t+1) = -1) &= \frac{1}{1 + e^{\beta F_i(t)}}
\end{aligned}
\tag{8.30}
$$

many times, you could calculate the average m_i of ξ_i,

$$
\begin{aligned}
m_i &= \frac{1}{1 + e^{-\beta F_i(t)}} - \frac{1}{1 + e^{\beta F_i(t)}} \\
&= \tanh(\frac{\beta}{2} F_i(t)).
\end{aligned}
\tag{8.31}
$$

Remember that, without thresholds,

$$F_i(t) = \sum_{j=1}^{n} T_{ij}\xi_j(t),\qquad\qquad(8.32)$$

It is now possible to replace $\xi_j(t)$ by its average m_j, and you obtain a deterministic network, where the state variables are averages of the state variables in the stochastic network.

The update rule for this network is

$$m_i(t+1) = \tanh\left(\frac{\beta}{2}\sum_{j=1}^{n} T_{ij}\frac{m_j(t)+1}{2}\right),\quad i = 1,\ldots,n.\qquad(8.33)$$

You can now replace the stochastic dynamics (8.30) with (8.33). This is not a stochastic network anymore, because it does not converge to an equilibrium distribution, but to an attractor with components $m_i, i = 1,\ldots,n$.

After replacing the network with another network, we will now also replace the averages of the product $\xi_i\xi_j$ with the product m_im_j of the averages. The learning procedure for the resulting network is called mean field annealing and operates as follows.

For the clamped average, set $m_i(0)$ equal to the clamped values, and choose random values for the other $m_i(0)$. Iterate (8.33) till an attractor is reached. Do this for successively increasing values of β, equivalent to decreasing the temperature. This way you obtain $< m_im_j >_c$. For the free running stage, do the same, without clamping. Modify the weights according to

$$\Delta T_{ij} = \eta(< m_im_j >_c - < m_im_j >_f),\quad i = 1,\ldots,n,\quad j = 1,\ldots,n.\qquad(8.34)$$

Again try to learn the exclusive-OR function with this learning rule. Compare the learning times with the deterministic rule (8.27) and the Boltzmann learning rule (8.29).

We have come full circle now, because the mean field dynamics in equation (8.33) are the same as those of a deterministic Hopfield network, with the function $\text{sgn}(x)$ replaced by $\tanh(\beta/2\,x)$. We have explored a stochastic network, speeded up its learning, and have arrived back at a deterministic network. The only difference is that the slope $\beta/2$ of the non-linearity varies during the learning stage.

The learning rules (8.27), (8.29), and (8.34) have another essential feature in common. They all use the correlations $x_i x_j$ between the neurons i and j to estimate the weigths T_{ij}. This idea was put forward by the neurobiologist Hebb [54] in the 1940's. By making the distinction between clamped and free running correlations, we have been able to apply it here to networks with hidden neurons.

Bibliography

[1] E. Aarts and J. Korst. *Simulated Annealing and Boltzmann Machines*. Wiley, New York, 1989.

[2] Ralph Abraham and Jerrold E. Marsden. *Foundations of Mechanics*. Benjamin, New York, 1967.

[3] Ralph H. Abraham and Christopher D. Shaw. *Dynamics, the Geometry of Behaviour*. Addison Wesley, Redwood City, California, 1992.

[4] Milton Abramowitz and Irene A. Stegun. *Handbook of Mathematical Functions*. Dover, New York, 1970.

[5] Yaser S. Abu-Mostafa and Jeannine-Marie St.Jacques. Information capacity of the Hopfield model. *IEEE Transactions on Information Theory*, IT-31(4):461–464, 1985.

[6] I. Aleksander and H. Morton. The logic of neural cognition. In R. Eckmiller, editor, *Proceedings of the International Symp. On Neural Networks For Sensory And Motor Systems, Neuss, Fed. Rep. Ger., March 1990*, pages 97–102, Amsterdam, 1990. Elsevier.

[7] Igor Aleksander. *Impossible Minds, My Neurons My Consciousness*. Imperial College Press, London, 1996.

[8] Igor Aleksander and Helen Morton. *Neurons and Symbols*. Chapman and Hall, London, 1993.

[9] Igor Aleksander and Helen Morton. *An Introduction to Neural Computing, 2nd. ed.* Thompson International, London, 1995.

[10] Daniel J. Amit. *Modeling Brain Function, The World of Attractor Neural Networks.* Cambridge University Press, Cambridge, 1989.

[11] Tom M. Apostol. *Mathematical Analysis.* Addison-Wesley, Reading, Massachusetts, 1974.

[12] Michael A. Arbib, editor. *The Handbook of Brain Theory and Neural Networks*, Cambridge, Massachussetts, 1995. MIT Press.

[13] Vladimir Igorevitch Arnol'd. *Geometrical Methods in the Theory of Ordinary Differential Equations.* Springer-Verlag, New York, 1983.

[14] D. K. Arrowsmith and C.M. Place. *An Introduction to Dynamical Systems.* Cambridge University Press, Cambridge, 1990.

[15] Pierre Baldi. Neural networks, acyclic orientations of the hypercube, and sets of orthogonal vectors. *Siam Journal on Discrete Mathematics*, 1(1):1–13, 1988.

[16] Pierre Baldi and Santosh S. Venkatesh. Number of stable points for spin-glasses and neural networks of higher orders. *Physical Review Letters*, 58(9):913–916, 1987.

[17] Christopher M. Bishop. *Neural Networks for Pattern Recognition.* Oxford University Press, Oxford, 1995.

[18] A. Braga and I. Aleksander. Geometrical treatment and statistical modeling of the distribution of patterns in the n-dimensional boolean space. *Pattern Recognition Letters*, 16(5):507–515, 1995.

[19] A. D. Bruce, E. J. Gardner, and D. J. Wallace. Dynamics and statistical mechanics of the Hopfield model. *Journal of Physics A: Mathematical and General*, 20:2909–2934, 1987.

[20] A. N. Chetaev. Some problems on markov chains arising in connection with the modelling of neural networks. *Russian Mathematical Surveys*, 31(4):77–87, 1976.

[21] A. N. Chetaev. *Nejronnye Seti i Tsepi Markova (Neural nets and Markov chains).* Nauka, Moskva, 1985.

[22] P.Y.K. Cheung, A. Ferrari, Ph. De Wilde, and G. Benyon-Tinker. A neural network processor - a vehicle for teaching system design. *IEE Proceedings-G*, 139(2):244–248, April 1992.

[23] Philip A. Chou. The capacity of the Kanerva associative memory. *IEEE Transactions on Information Theory*, 35(2):281–298, March 1989.

[24] Barry A. Cipra. An introduction to the Ising model. *American Mathematical Monthly*, 94(10):937–959, 1987.

[25] John W. Clark. Statistical mechanics of neural networks. *Physics Reports*, 158(2):91–157, 1988.

[26] D. R. Cox and H. D. Miller. *The Theory of Stochastic Processes*. Chapman and Hall, London, 1984.

[27] Philippe De Wilde. A Marquardt learning algorithm for neural networks. In A. M. Barbé, editor, *Proceedings of the Tenth Symposium on Information Theory in the Benelux*, pages 51–57, Enschede, The Netherlands, 1989. Werkgemeenschap voor Informatie- en Communicatietheorie.

[28] Philippe De Wilde. Time and area requirements for a hybrid learning algorithm. In M. Novak and E. Pelikán, editors, *Proceedings of the International Symposium on Neural Networks and Neural Computing NEURONET'90*, pages 354–356, Prague, 1990. Czechoslovak Academy of Sciences.

[29] Philippe De Wilde. Conditions for active states in neural networks. In A. Holden and V. Kryukov, editors, *Neurocomputers and Attention*, pages 695–700, Manchester, 1991. Manchester University Press.

[30] Philippe De Wilde. Class of Hamiltonian neural networks. *Physical Review E*, 47(2):1392–1396, February 1993.

[31] Philippe De Wilde. Reduction of representations and the modelling of consciousness. In Harald Hüning et al., editor, *Aachener Beiträge zur Informatik, Band 3*, pages 40–43, Aachen, 1993. Verlag der Augustinus Buchhandlung.

[32] Philippe De Wilde. Physical and linguistic problems in the modelling of consciousness by neural networks. In J. Mira and F. Sandoval, editors, *Lecture Notes in Computer Science 930*, pages 584–588. Springer Verlag, Berlin, 1995.

[33] Philippe De Wilde. The size of the diagonal elements in neural networks. *Neural Networks*, 1997. To be published.

[34] Philippe De Wilde and A.M.C.-L. Ho. Symmetries of general feedforward neural networks and equivalent classification tasks. In I. Aleksander and J. Taylor, editors, *Artificial Neural Networks II, Proceedings of the International Conference on Artificial Neural Networks*, Amsterdam, 1992. North-Holland.

[35] R. O. Duda and P. E. Hart. *Pattern Classification and Scene Analysis*. Wiley, New York, 1973.

[36] J. C. Eccles. *The Physiology of Synapses*. Springer Verlag, Berlin, 1964.

[37] John C. Eccles. A unitary hypothesis of mind brain interaction in the cerebral cortex. *Proceedings of the Royal Society of London Series B - Biological Sciences*, 240(1299):433–451, 1990.

[38] John C. Eccles. Evolution of consciousness. *Proceedings of the National Academy of Sciences of the USA*, 89(16):7320–7324, 1992.

[39] G. B. Ermentrout and J. D. Cowan. Large scale spacially organized activity in neural nets. *SIAM Journal of Applied Mathematics*, 38(1):1–21, 1980.

[40] Wai-Chi Fang, Bing J. Shen, Oscal T.-C. Chen, and Joongho Choi. A VLSI neural processor for image data compression using self-organization networks. *IEEE Transactions on Neural Networks*, 3(3):506–518, 1992.

[41] W. J. Freeman. Neural networks and chaos. *Journal of Theoretical Biology*, 171(1):13–18, 1994.

[42] Stephen I. Gallant. *Neural Network Learning and Expert Systems*. MIT Press, Cambridge, Massashusetts, 1993.

[43] Elizabeth Gardner. The space of interactions in neural networks. *Journal of Physics A: Mathematical and General*, 21(1):257–270, 1988.

[44] James Gleick. *Chaos.* Viking, New York, 1987.

[45] Eric Goles and Servet Martínez. *Neural and Automata Networks.* Kluwer, Dordrecht, 1990.

[46] Rafael C. Gonzalez and Richard E. Woods. *Digital Image Processing.* Addison-Wesley, Reading, Massachusetts, 1993.

[47] I. S. Gradshteyn and I. M. Ryzhik. *Table of Integrals, Series, and Products.* Academic Press, New York, 1980.

[48] Hans P. Graf, Lawrence D. Jackel, and Wayne E. Hubbard. VLSI implementation of a neural network model. *Computer*, 21(3):41–49, 1988.

[49] Ronald L. Graham, Donald E. Knuth, and Oren Patashnik. *Concrete Mathematics.* Addison-Wesley, Reading, Massachusetts, 1994.

[50] Stephen Grossberg. *The adaptive brain*, volume I & II. North-Holland, Amsterdam, 1986.

[51] Hermann Haken. *Synergetics.* Springer-Verlag, Berlin, 1983.

[52] Michael A. Harrison. *Introduction to Switching and Automata Theory.* McGraw-Hill, New York, 1965.

[53] Simon Haykin. *Neural Networks, a Comprehensive Foundation.* Macmillan, New York, 1994.

[54] D. O. Hebb. *The Organization of Behaviour.* Wiley, New York, 1949.

[55] Robert Hecht-Nielsen. Neurocomputing: picking the human brain. *IEEE Spectrum*, 25(3):36–41, 1988.

[56] John Hertz, Anders Krogh, and Richard G. Palmer. *Introduction to the Theory of Neural Computation.* Santa Fe Institute Studies in the Sciences of Complexity. Addison-Wesley, Redwood City, California, 1991.

[57] A.M.C.-L. Ho and Philippe De Wilde. General transient length upper bound for recurrent neural networks. In J. Mira and F. Sandoval, editors, *Lecture Notes in Computer Science 930*, pages 202–208. Springer Verlag, Berlin, 1995.

[58] A. L. Hodgkin and A. F. Huxley. A quantitative description of current and its application to conduction and excitation in nerve. *Journal of Physiology*, 117:500–544, 1952.

[59] John E. Hopcroft and Jeffey D. Ullman. *Introduction to Automata Theory, Languages, and Computation*. Addison-Wesley, Reading, Massachusetts, 1979.

[60] John J. Hopfield. Neural networks and physical systems with emergent collective computational abilities. *Proceedings of the National Academy of Sciences of the USA*, 79:2554–2558, 1982.

[61] John J. Hopfield and David W. Tank. Computing with neural circuits: a model. *Science*, 233(4764):625–633, 1986.

[62] Eugene Isaakson and Herbert Bishop Keller. *Analysis of Numerical Methods*. John Wiley & Sons, New York, 1966.

[63] Yves Kamp and Martin Hasler. *Recursive Neural Networks for Associative Memory*. Wiley, Chichester, 1990.

[64] I. Kanter and H. Sompolinsky. Associative recall of memory without errors. *Physical Review A*, 35(1):380–392, 1987.

[65] Maria Karapataki and Philippe De Wilde. The hopfield network applied to blood vessel detection in angiograms. *Medical & Biological Engineering & Computing*, 1997. To be published.

[66] James D. Keeler. Information capacity of outer-product neural networks. *Physics Letters A*, 124(1/2):53–58, 1987.

[67] Teuvo Kohonen. An introduction to neural computing. *Neural Networks*, 1(1):3–16, 1988.

[68] János Komlós and Ramamohan Paturi. Convergence results in an associative memory model. *Neural Networks*, 1(3):239–250, 1988.

[69] Fukumi Kozato and Philippe De Wilde. How neural networks help rule-based problem solving. In T. Kohonen, K. Mäkisara, O. Simula, and J. Kangas, editors, *Artificial Neural Networks, Proceedings of the 1991 International Conference on Artificial Neural Networks*, pages 465–470, Amsterdam, 1991. North-Holland.

[70] Fukumi Kozato and Philippe De Wilde. A probabilistic rule-based system in artificial neural networks. In *Proceedings of the Second International Conference on Artificial Neural Networks*, London, 1991. IEE.

[71] V. I. Kryukov, G. N. Borisyuk, R. M. Borisyuk, A. B. Kirillov, and E. I. Kovalenko. *The Metastable and Unstable States in the Brain (in Russian)*. Academy of Sciences of the USSR, Pushchino, 1986.

[72] Solomon Lefschetz. *Differential equations: geometric theory*. Wiley, New York, 1963.

[73] Jian Hua Li, Anthony N. Michel, and Wolfgang Porod. Qualitative analysis and synthesis of a class of neural networks. *IEEE Transactions on Circuits and Systems*, CAS-35(8):976–985, 1988.

[74] Ralph Linsker. Self-organization in a perceptual network. *Computer*, pages 105–117, 1988.

[75] C. M. Markus, F. R. Waugh, and R. M. Westervelt. Associative memory in an analog iterated-map neural network. *Physical Review A*, 41(6):3355–3364, 1990.

[76] E. Marom. Associative memory neural networks with concatenated vectors and nonzero diagonal terms. *Neural Networks*, 3(3):311–318, 1990.

[77] James L. McClelland and David E. Rumelhart. *Explorations in Parallel Distributed Processing: A Handbook of Models, Programs, and Exercises*. MIT Press, Cambridge, Massachusetts, 1988.

[78] James L. McClelland, David E. Rumelhart, and the PDP Research Group. *Parallel Distributed Processing: Explorations in*

the Microstructure of Cognition, Volume 2, Psychological and Biological models. MIT Press, Cambridge, Massachusetts, 1986.

[79] Warren S. McCulloch. *Embodiments of Mind.* MIT Press, Cambridge, Massachusetts, 1989.

[80] Robert J. McEliece, Edward C. Posner, Eugene R. Rodemich, and Santosh S. Venkatesh. The capacity of the Hopfield associative memory. *IEEE Transactions on Information Theory*, IT-33(4):461–482, 1987.

[81] Carver Mead. *Analog VLSI and Neural Systems.* Addison-Wesley, Reading, Massachusetts, 1989.

[82] Marc Mézard, Giorgio Parisi, and Miguel Angel Virasoro. *Spin Glass Theory and Beyond.* World Scientific, Singapore, 1987.

[83] Anthony N. Michel, Jay A. Farrell, and Wolfgang Porod. Qualitative analysis of neural networks. *IEEE Transactions on Circuits and Systems*, CAS-36(2):229–243, 1989.

[84] John Milton. *Dynamics of Small Neural Populations.* CRM Monograph Series Vol.7. American Mathematical Society, Providence, Rhode Island, 1996.

[85] Marvin Minsky and Seymour Papert. *Perceptrons : an introduction to computational geometry.* MIT Press, Cambridge, Massachusetts, 1988.

[86] J. E. Moreira, F. W. S. Lima, and J. S. Andrade. Controlling chaos by pinning neurons in a neural network. *Physical Review E*, 52(3A):R2129–R2123, 1995.

[87] Saburo Muroga. *Threshold Logic and its Applications.* Wiley, New York, 1971.

[88] I. P. Natanson. *Constructive function theory*, volume 1–3. Ungar, New York, 1965.

[89] Reza Nekovei and Ying Sun. Back-propagation network and its configuration for blood vessel detection in angiograms. *IEEE Transactions on Neural Networks*, 6(1):64–72, 1995.

[90] Charles M. Newman. Memory capacity in neural network models: Rigorous lower bounds. *Neural Networks*, 1(3):223–238, 1988.

[91] Athanasios Papoulis. *Probability, Random Variables, and Stochastic Processes.* McGraw-Hill, New York, 1981.

[92] Thomas S. Parker and Leon O. Chua. Chaos: A tutorial for engineers. *Proceedings of the IEEE*, 75(8):982–1008, 1987.

[93] François Robert. *Discrete Iterations: a Metric Study.* Springer-Verlag, Berlin, 1986.

[94] David E. Rumelhart, James L. McClelland, and the PDP Research Group. *Parallel Distributed Processing: Explorations in the Microstructure of Cognition, Volume 1, Foundations.* MIT Press, Cambridge, Massachusetts, 1986.

[95] Yakov Grigor'evich Sinai. *Topics in Ergodic Theory.* Princeton University Press, Princeton, New Jersey, 1994.

[96] John F. Sowa. *Conceptual Structures.* Addison Weslay, Reading, Massachusetts, 1984.

[97] Murray R. Spiegel. *Advanced Calculus.* McGraw-Hill, New York, 1974.

[98] Gilbert Strang. *Linear algebra and its applications.* Harcourt Brace Jovanovich, San Diego, 1988.

[99] Dawid W. Tank and John J. Hopfield. Simple "neural" optimization networks: An A/D converter, signal decision circuit, and a linear programming circuit. *IEEE Transactions on Circuits and Systems*, CAS-33(5):533–541, 1986.

[100] Santosh S. Venkatesh. Robustness in neural computation: Random graphs and sparsity. *IEEE Transactions on Information Theory*, 38(3):1114–1119, 1992.

[101] Ferdinand Verhulst. *Nonlinear Differential Equations and Dynamical Systems.* Springer-Verlag, Berlin, 1990.

[102] DeLiang Wang. Emergent synchrony in locally coupled oscillators. *IEEE Transactions on Neural Networks*, 6(4):941–948, 1995.

[103] K. Y. M. Wong and D. Sherrington. Theory of associative memory in randomly connected Boolean neural networks. *Journal of Physics A: Mathematical and General*, 22(12):2233–2263, 1989.

Index